Edges of Exposure

EXPERIMENTAL FUTURES · Technological Lives, Scientific Arts, Anthropological Voices *A series edited by Michael M. J. Fischer and Joseph Dumit*

Edges of Exposure

Toxicology and the Problem of Capacity in Postcolonial Senegal · NOÉMI TOUSIGNANT

Duke University Press   Durham and London   2018

Interior designed by Courtney Leigh Baker
Typeset in Minion Pro and Avenir by Copperline Book Services

Library of Congress Cataloging-in-Publication Data
Names: Tousignant, Noémi, [date] author.
Title: Edges of exposure : toxicology and the problem of
capacity in postcolonial Senegal / Noémi Tousignant.
Description: Durham : Duke University Press, 2018. |
Series: Experimental futures | Includes bibliographical
references and index.
Identifiers: LCCN 2017049286 (print)
LCCN 2017054712 (ebook)
ISBN 9780822371724 (ebook)
ISBN 9780822371137 (hardcover : alk. paper)
ISBN 9780822371243 (pbk. : alk. paper)
Subjects: LCSH: Environmental toxicology—Senegal. |
Heavy metals—Toxicology—Research—Senegal. |
Environmental justice—Senegal.
Classification: LCC RA1226 (ebook) |
LCC RA1226 .T66 2018 (print) | DDC 615.9/0209663—dc23
LC record available at https://lccn.loc.gov/2017049286

Cover art: Toxicology teaching laboratory.
Photo by Noémi Tousignant, Dakar, 2010.

To the memory of
Binta "Yaye Sall" Kadame Gueye,
of Mame Khady Gueye and Maya Gueye,
and of Jeannette Baker Tousignant

CONTENTS

ACKNOWLEDGMENTS

I had a lot of help making my way to this book. Much of this help was at once intellectual and affective, at once practical and vital, and exceeded what I asked or hoped for. Such unruly generosity cannot be fully acknowledged.

Family gave much more than emotional support. They helped me navigate my field-site and the writing path. Moussa Gueye made me understand the fragility of hope and opportunity for the "adjustment generation" in Senegal while giving strength to our future. Avi Daouda and Clara Binta "Yaye Sall Kadame" Tousignant Gueye brought brightness and vitality to my work. Sherry Simon and Michel Tousignant know everything, as scholars, about writing a book—the pleasures, labours, hesitations, and rewards. As parents, they made it seem a unique achievement. Shirley Berk Simon, with avid curiosity and grandmotherly pride, drew me toward the goal. Eleanor Wachtel, great friend to writers of all sorts, lured me into the world, as did Tzippy Corber, with her wide-open ears and eyes, and Tobie Tousignant, always graceful behind the wheel and with a racket, whose trademark easiness prepared me for the last *ligne droite*. The Gueye family in Dakar opened many kinds of doors for me and made me feel that my place was inside, especially Assane Gueye, Daouda and Aïda Gueye, as well as Bouba Mane, with "Ta" Ndumbe, who opened the very first doors, domestic and professional. "Tata" Maria Hernandez joined me in laughter and ethnography. The Tingays—Clare, Ant, Hugh, Sophia, and Rebecca—gave me a home and a family in London.

My "informants" provided so much more than information. Some became friends; all gave generously of their time, ideas, and connections, helping me make sense of their lives and hopes. I am especially grateful to Pr. Amadou Diouf and Dr. Dogo Seck for trusting me with their institutions and histories, and to Pr. Mamadou Fall, Dr. Baba Gadji, Dr. Babacar Niane, and Dr. Marieme Mbaye Sene, who guided me through the spaces and histories of the Centre Anti-Poison, UCAD's Laboratoire de toxicologie et d'hydrologie, and CERES-Locustox. Thanks to the dozens of staff members of these institutions and others who took time to speak with me, with special thanks to Doudou Ba and Mounirou Ciss, the *grands* who sparked my interest in Senegalese toxicology. Pr. Amadou Moctar Dièye, Charles Becker, and Myron Echenberg helped me take my first steps in Dakar, while Bernard Taverne and Alice Desclaux gave me space to write and introduced me to new networks.

I owe a very great debt to Wenzel Geissler, for his insight and generosity, for his reluctant but highly effective "management," and above all for his high expectations—of me, but also of scholarship on African science in general and of African scientists themselves—which have made this a much better book. He, Ruth Prince, and Guillaume Lachenal read drafts and gave freely of good ideas, good advice, and good hospitality. Two anonymous readers asked the questions that helped me write what I really meant.

I have been fortunate to work with colleagues in London, Cambridge, Montreal, and in the far-flung MEREAF project who have opened to me both exceptionally sharp minds and wonderfully warm hearts: Gemma Aellah, Dörte Bemme, Rob Boddice, Hannah Brown, Martha Chinouya, Jennifer Cuffe, Pierre-Marie David, Rene Gerrets, Hannah Gilbert, Nick Hultin, Lauren Hutchinson, Loes Knaapen, Christos Lynteris, Peter Mangesho, Aïssatou Mbodj-Pouye, Karen McAllister, David Meren, Pierre Minn, Anne-Marie Moulin, Raul Necochea, Stephanie Olsen, Ashley Ouvrier, Joseph Owona Ntsama, David Reubi, Rémy Rouillard, and Thomas Schlich. I benefited especially from closer collaborations with Uli Beisel, Alice Desclaux, Ann Kelly, John Manton, Branwyn Poleykett, and again Wenzel, Ruth, and Guillaume. Laurence Monnais and George Weisz taught me much and stuck with me, and I have kept learning as a colleague.

I drew ideas and inspiration from shorter conversations, with audiences of earlier versions of this work and with guests in events and publications around themes connected to it. I have gained much from crossing paths with Ash Amin, Casper Andersen, Warwick Anderson, Johanna Crane, Filip De Boeck, Ferdinand De Jong, Damien Droney, David Dunne, Steve Feierman, Tamara Giles-Vernick, Melissa Graboyes, Jeremy Greene, John Harrington, Penny Harvey, Gabrielle Hecht, Sarah Hodges, Nancy R. Hunt, Katie Kilroy-Marac, Harun Küçük, Johan Lagae, Janelle Lamoreaux, Julie Livingston, Doreen Massey, Ramah McKay, Marissa Mika, Henrietta Moore, Projit Mukharji, Michelle Murphy, Iruka Okeke, Anne Pollock, Mathieu Quet, Joanna Radin, Peter Redfield, Tobias Rees, Susan Reynolds Whyte, François Richards, Simon Schaffer, Nick Shapiro, Kavita Sivaramakrishnan, Noelle Sullivan, Marlee Tichenor, Megan Vaughan, John H. Warner, and Claire Wendland.

Linda Amarfio, Madeline Watt, and Laura Cousens provided invaluable administrative support. Work toward this book was mainly funded by the Leverhulme Trust (Research Leadership Award, Geissler, F02 116D), while connected projects and conversations were also made possible by research and conference support from the British Economic and Social Research Council (RES-360-25-0032), the French Agence Nationale de Recherche Scientifique (ANR-AA-ORA-032), the Canadian Social Science and Research Council, and the Wellcome Trust (WT 092699MF and further conference support). It has been a great privilege to work with Courtney Berger, Sandra Korn, and others at Duke University Press.

Introduction · Poisons and Unprotection in Africa

In early 2008, astonishing levels of lead were detected in the soil and blood-streams of the community of Ngagne Diaw. Over a prior four-month span, eighteen young children had died there, in a neighborhood of about 950 residents nestled between the coast and the main road leading out of Dakar, the capital city of Senegal in West Africa. Other children suffered convulsions, vomiting, brain inflammation, and loss of concentration and muscle coordination. Some siblings of the deceased children were found to have blood-lead concentrations above the threshold considered to be fatal. Investigations traced this exposure to a recent surge in the price of lead purchased by Indian entrepreneurs. Residents of Ngagne Diaw had long broken and burned used lead-acid car batteries (ULABS) to scrape out lead for fishing weights. In 2005, however, this recuperation activity intensified. Battery debris and lead scraps piled up in and around homes. Toxic lead dust settled on the ground, walls, and floors. It touched skin and was inhaled and ingested.[1]

What can this tragedy tell us about poisons in Africa, and in particular about missing and possible protections? By 2010, when I came to Dakar to study the contemporary history of toxicology, various interpretations were taking shape. One was argued by Adama Fall, a Senegalese lawyer, in a recent prize-winning plea for this case in an international human rights law competition.[2] Fall set up a striking parallel between the killing of children by lead and the earlier shooting of thirty-eight West African soldiers by the French colonial army in the nearby military camp of Thiaroye in December 1944. This was not simply to point out that the place was prone to tragedy. The massacre of Thiaroye, a topic of well-known films as well as poems, plays, and a novel, has become symbolic of the hypocrisy of late-colonial promises of citizenship.[3] At the very time when France was recognizing its colony's services to the nation and rights to political participation, the African soldiers' demands for decent working conditions and back pay were violently repressed. Seventy years later, a stone's throw from the soldiers' graves, the state had again, Fall implied, betrayed its (potential) citizens. Fall accused the Senegalese state of a series of specific failures: to enforce a long list of national laws and ratified international conventions on worker and environmental protection, and on toxic waste; to effectively regulate and prosecute the guilty transnational firm; and to care for and compensate the poisoned. A proposed plan to relocate Ngagne Diaw's inhabitants had even triggered rumors of a plot by state authorities to grab valuable land in the bottleneck of the densely inhabited Cap-Vert Peninsula. By locating the poisoning in a longer history of state betrayal, Fall echoes readings of other toxic tragedies in Africa, notably the illegal dumping of waste in the city of Abidjan, Côte d'Ivoire, in 2006. These feature a predatory, or at least powerless, state exposing its population to literally poisonous global capital, which inflicts the riskiest forms of extraction and disposal on the cheapest, least-protected lives.[4]

This first interpretation, then, focuses on *exposure*. In the management of the contamination, a second interpretation took form, which instead cast the crisis as a problem of technical capacity. The Blacksmith Institute, a US-based NGO describing itself as "dedicated to solving life-threatening pollution issues in low- and middle-income countries,"[5] was the first institution called in to help in Ngagne Diaw.[6] Although the NGO (since renamed Pure Earth) calls attention to the synergy between toxic risk and economic vulnerability (captured, a few years after the decontamination of Ngagne Diaw, by the expression "the poisoned poor"), its focus is pragmatic. It frames poisoning as a humanitarian crisis, requiring immediate, mobile, and minimal yet lifesaving protections.[7] In Ngagne Diaw, Blacksmith provided expertise and financing for soil removal, house-to-house cleaning, and an awareness-raising campaign on the dangers of lead. It also provided a portable testing apparatus to measure initial and dropping lead concentrations in soil and blood, thereby obtaining proof of the operation's success in averting the "imminent danger" of epidemic poisoning.[8] Thus Blacksmith presented its intervention as bridging vital gaps in Senegalese state capacity to monitor and eliminate deadly exposures.

Those who carried out the activities supported by Blacksmith and by the World Health Organization (WHO) were experts and technicians working for the Senegalese state. Notably, a small team from the recently established Centre Anti-Poison (Poison Control Center, CAP) was the first to diagnose elevated blood-lead levels. The CAP staff then assisted the Blacksmith and WHO teams in confirming the poisoning and was then put in charge of screening, monitoring, education, and the organization of the drug supply for chelation therapy (to remove lead from the body). Led by toxicologist Amadou Diouf, who was also head of the toxicology department of Dakar's Université Cheikh Anta Diop (UCAD), this team's commitment to detecting and managing poison in Senegal was not limited to the time and place of the crisis of Ngagne Diaw—a "toxic hotspot" in Blacksmith's vocabulary. It was part of a much longer history of efforts to measure and monitor toxic threats in Senegal.

From Diouf and his team's perspective, the horizons of missing and possible protections from poison extended well beyond Ngagne Diaw. Since the 1970s, toxicologists at the university (then named the Université de Dakar) have worked to track down toxic traces in Senegalese bodies and environments, and called for the expansion and routinization of surveys and testing. The head of the university's toxicology unit proposed to create a national poison control center as early as 1973. As money for research and testing came and went, as equipment arrived and broke down, as the plausibility of expansive, regular control faded, toxicologists continued to investigate, even if sometimes on tiny scales, the presence of poisons in Senegal. The very existence of a poison control center, if only as a modest staff and operating budget, at the time the tragedy struck in 2008 owed much to Diouf's tenacious lobbying over the previous years. Diouf's preexisting connections to a private medical laboratory in Dakar, as well as with Blacksmith, also shaped the response to Ngagne Diaw. The private lab helped Diouf get the first tests for lead done on blood drawn from siblings of the deceased children (its staff performed some routine tests on the blood, then shipped samples to France for other tests, all for free). Indeed, Diouf had relied on these connections before to obtain free testing for a study of exposure to lead in a tiny sample of car mechanics and ULAB recyclers in Dakar—this experience, some colleagues suggested, helped him guess the cause of deaths in Ngagne Diaw. By 2010, the CAP team was working, from a half-finished building and without any laboratory equipment, to take poison control beyond crisis control and into regular, long-term surveillance and response services such as the collection of epidemiological and incident data and a 24-7 poison helpline.

For Senegalese toxicologists and their colleagues,[9] the pasts and futures of Ngagne Diaw are not only of cumulative exposure, as in Fall's plea, or of a static gap in capacity, as Blacksmith sought to bridge. They are of working, succeeding, and failing to gain, keep, and stretch the material and institutional capacity needed to detect and define toxic risks in the country.[10] This history of *struggle for capacity* is what I describe in this book. The rhythms of this struggle have been intermittently set in motion by investments in Senegalese scientific research, in higher education, and, occasionally, in the monitoring of toxic contamination. More often these rhythms have been stilled and interrupted by stagnating budgets, the end of project funding, the breakdown of equipment, and the wait for an overseas trip. Yet pushing into and against broken rhythms of funding, supply, and repair, toxicologists have also fought to extend and hold together fragments of testing capacity and knowledge. They have sought to set the cadence of acts of detection, surveying, and surveillance into more regular, continuous, and cumulative patterns. This struggle is not one of heroic self-sacrifice for the public good; toxicologists in Senegal have, like most scientists anywhere, pursued capacity as a condition of professional survival and success. Still, they have done so as *public* scientists, defined as such by their funding and institutions, and also as practitioners of a set of techniques and expertise that, historically, has become central to how modern industrial societies protect their publics from collective toxic risks.[11] Their pursuit of capacity (and narration of this pursuit) has thus attempted to tie professional ambition to public service and protection.

In this, their success has been partial at best. The majority of their studies have been modest in scope and scale, revealing points of contamination— for example, the presence of pesticides and aflatoxins (the toxic metabolites of some strains of fungus that grow on foodstuffs, particularly under poor storage conditions) in some foods—that have not been linked up to more extensive surveys, regular monitoring, or regulatory action. This work has probably been more effective in obtaining publications and promotions for toxicologists than any real protections for the Senegalese public. Ultimately, the struggle for toxicological capacity seems largely futile, unable to generate protective knowledge other than as fragments, hopes, and fictions.[12] Still, these fragments count; they map the partial contours of a "landscape of exposure,"[13] pointing not merely to the absence of capacity and protection but to its edges and missed possibilities, where knowable toxicities circulate, uncaptured by analytical equipment, epidemiological surveys, or monitoring routines.

Following toxicologists through their three main institutions in Senegal—a public university laboratory, an ecotoxicological project/center, and a national poison control center—this book weaves together an account of intermittent and insufficient investments in toxicological capacity with fine-grained descriptions of how scientists have kept equipment, labs, projects, and careers going. Its main focus is on what "good science" has meant—in practice, memory, goals, and dreams—to chronically underfunded and ill-equipped scientists. In this, protection from poison figures more as a form of moral imagination (or fiction), which gives value to fragmentary and sought-after capacity, than as a fully articulated vision of how enhanced capacity might initiate and feed into a denser and more effective network of mechanisms of prevention and control. Indeed, neither I, nor toxicologists in Senegal, suggest that better-equipped laboratories and institutions would automatically, or directly, translate into better-protected populations. Avoiding exposure requires many forms of protection, from expanded choices about where to work and live to a variety of types of regulatory investigation and action. Yet the detection of toxicities seems a crucial step in making contamination a topic of public debate and public protection.

It is true that no one is fully free or protected from exposure. Toxicology in Senegal (or other low-resource settings in Africa and the Global South) may not be exceptional in its "powerlessness" to control risk.[14] The world we now live in, some say, is toxic; our bodies are all a bit synthetic.[15] Seeping across social, spatial, and biological lines, omnipresent toxicity is, in Ulrich Beck's "risk society," part and parcel of the inherent risk of late-modern society; its uncontrollability manifests the limits of scientific expertise.[16] Industrial capitalism generates not only risk, David Pellow adds, but also inequality.[17] Economic and environmental vulnerabilities intersect in the uneven distribution—on both national and global scales—of the toxic burdens of progress and growth. Yet whether emphasizing the inevitability of exposure or its uneven distribution, there is a widespread tendency to take for granted that at least *minimal* acts of toxicological detection and protection are routinely provided to residents of the Global North and are largely absent in the Global South.[18]

To delineate gaps in toxicological capacity is to acknowledge that opportunities to protect from (and to politicize) toxic risk are *withheld.* There is an excellent literature on toxicology in higher-resource settings, especially in the United States, exploring how and why potentially protective knowledge has been obstructed and obfuscated, and how scientists have, in some cases, fought against these limitations.[19] Yet very little scholarly attention is paid to toxicology and toxicologists under more extreme conditions of material scarcity, dependence, and uncertainty. In other words, the overlapping geographies of environmental and scientific dispossession, where "the poisoned poor," in Blacksmith's words, meet, in Africa, what Paulin Hountondji has called "impoverished science,"[20] are largely unexamined. This is the space through which this book moves. It focuses, in its details, on what (un)protection means to scientists' own understandings of capacity, identity, success, and service. On a more general level, however, it is also a plea to invest—for the sake of public health, environmental control, and public debate—in toxicologists' capacity to reveal, measure, map, and keep tabs on the presence of otherwise invisible forms of contamination and exposure in Africa, or elsewhere such capacity is inadequate. I also want to give recognition not only to the futility but also to the persistence, energy, and hopefulness of toxicologists' pursuit, in Senegal, of toxicology as a public and protective science.

AFRICAN MAPS OF EXPOSURE

Poisons in Africa raised scandal before the tragedy in Ngagne Diaw. Three occurrences in particular have prompted commentary on Africans' extreme exposure to the risks generated by a globalizing economy. An early wave of protest arose when, in the late 1980s, the story broke that hazardous industrial waste was being exported from wealthy economies to West Africa.[21] The patterns of environmental racism—siting toxic production and waste near the dispossessed and discriminated—that were under protest in the United States[22] seemed to be going global.[23] A few years earlier, in 1984, a lethal toxic leak at the American-owned pesticide plant in Bhopal, India, was interpreted as a manifestation of the literally poisonous effects of trade liberalization in an unequal world, facilitated by poverty and unchecked by adequate mechanisms of accountability, regulation, and wealth distribution.[24] Yet waste dumping in Africa, labeled "toxic terrorism,"[25] was also a specific reminder, in the words of journalist Sam Omatseye, "of what Europe has always thought of Africa: A Wasteland. And the people who live there, waste beings."[26]

The second scandal emerged around a leaked memo, signed in 1991 by Lawrence Summers, then chief economist at the World Bank. The memo defended (in jest, Summers claimed) the migration of "dirty industries" to developing countries, citing Africa specifically as "under-polluted" but also as less likely to resist with demands for "a clean environment for aesthetic and health reasons."[27] Scholars seized on the memo as an exceptionally blunt expression of the logic underlying the distribution of toxic risk, and of how this logic exposed Africa in particular. For Rob Nixon, the memo "triply" dismisses Africans: as political agents, as victims of pollution, and as environmentalists.[28] James Ferguson presents it as a "raw form" of the reasoning by which the World Bank justified structural adjustment programs (SAPs) in Africa, suspending "social and moral values" as (potentially protective) buffers of economic rationality.[29] In the introduction to her ethnography of cancer care in Botswana, Julie Livingston cites the memo to illustrate how the prevailing model of epidemiological transition has posited Africans as "biologically simple publics" whose pretransition bodies—afflicted by "infectious disease, fertility and malnutrition"—are unlikely to register toxic effects, especially delayed ones like cancer.[30] Leaked shortly before the implementation of the Basel Convention—an international agreement on the transboundary circulation of hazardous waste adopted in 1989—the memo seemed to warn that regulatory responses would not be enough to stop the powerful forces driving toxic redistribution.[31]

Sure enough, the densification of this international regulatory framework on toxics (the Basel Convention was followed by the Rotterdam and Stockholm Conventions, adopted in 1998 and 2001, respectively)[32] failed to prevent the disposal of dripping drums of toxic caustic sludge around the city of Abidjan, Côte d'Ivoire, in 2006. The official toll was of fifteen to seventeen deaths, and more than 100,000 cases of "nausea, headaches, breathing difficulties, stinging eyes and burning skin."[33] Tracing the sludge to a Dutch commodity-trading company and an Ivoirian company's offer of cheap disposal, an Amnesty International and Greenpeace investigation casts this exposure as a "story of corporate crime, human rights abuse and governments' failure to protect people and the environment."[34] This failure of protection has been analyzed in explicitly postcolonial terms. The legal scholar Lassana Koné places the illegal dumping—along with the more insidious exports of waste for "recycling," which international mechanisms have struggled to regulate—under the label of "toxic colonialism."[35] As an example of how empire, as a persistent process of ruination, exerts aftereffects, Ann Laura Stoler calls the Abidjan sludge "toxic debris."[36] For Alex Means, the tragedy exemplifies the Ivoirian state's "toxic sovereignty," which, following Michael Hardt and Antonio Negri, is reduced to an "emergency apparatus" that functions only to facilitate the smooth circulation of capital.[37] Set against post-independence expectations in Africa, especially in Côte d'Ivoire, of economic and political emergence and future global convergence,[38] the "stinking toxic waste," writes Sarah Lincoln, materialized "the gap between postcolonial expectation and postmodern disillusionment."[39] Véronique Tadjo, in a collection of texts for the fiftieth anniversary of African independence, aptly expresses the imbrication of contamination with lost hopes of sovereignty and accountability:

> Independence. Liberating ourselves from fatalism and wasted destinies. . . . I haven't even spoken of the toxic waste dump, still alive, still active in the heart of the city. This acrid smell in the air, it is the poison they force us to breathe in. . . . We must refuse, rise up against dereliction. But who are our masters anyhow? Who are they to not pity those they govern. . . . Our greatest struggle, our real independence now is way beyond the squabbling of politicians who wreck our existence. It is what we will leave behind us that matters, what we do of our present that counts.[40]

The framing of poisons in Africa has thus drawn on a strong association between toxicity and *waste*, both as literal waste that is dumped, as an external, material assault on Africans, and as symbolic of the continent's superfluity in the global political-economic order, that is, of Africa *as* waste.[41] This is a useful starting point for thinking about the cumulative and intersecting vulnerabilities—biological, economic, political—through which exposure is amplified. It also links failures of protection both to the political-economic constraints facing African states and to the continent's place in the global imagination: as a dumping ground; as epidemiologically, environmentally, and technologically not yet modern; as cheap and unregulated; and as in need of rational economic solutions. This framing, however, largely eludes two issues that are of great concern to toxicologists in Africa, and which therefore inform this book.

The first is what Rob Nixon calls the "predicaments of apprehension."[42] Not only is waste symbolically charged, but it is also, as in the cases that underpin the analyses cited previously, explicit and perceptible as odor, leaky drums, and fatalities. More often, however, poison is a hidden presence, in traces and residues, while toxicity is sublethal, its effects subtle and delayed. This "slow-motion toxicity," as Nixon calls it, poses the "challenge, at once imaginative and scientific, of giving the unapparent a materiality upon which we can act."[43] Yet, as he points out, the very people and spaces most "exposed to the force field of slow violence are abandoned to sporadic science at best and usually no science at all."[44] As Julie Livingston and Gabrielle Hecht have shown, the conceptual exclusion of Africa from epidemiological and "nuclear" modernity has kept (potential) exposures invisible. That Africans are seen as not yet vulnerable to a disease (cancer) defined as a pathology of "civilization," and African uranium miners seen as not performing "nuclear" work, has justified the absence of research on patterns of cancer causation and prevalence as well as on radiation as an occupational risk.[45] Toxicologists in Senegal, and elsewhere in Africa, have taken up Nixon's challenge and worked against this toxic invisibility to reveal a finer-grained, more varied and complex map of contamination that stretches beyond dumped waste and migrating industries to follow foodstuffs, riverways, and bloodstreams. Yet this map is also marked by the limits of and on toxicologists' capacity; by the constraints posed by the low status of poisoning in national and global health and environmental agendas and by the poor state of their laboratories; and by the restricted size of their sample sets and range of analytical tests and number of studies they can perform. Toxicologists thus highlight both the work it takes to make toxicity visible and the obstacles in their path, hinting—in their partial results, their calls for more testing, their complaints about incapacity and dysfunctional regulation—at the large swathes of invisible toxicity that lie beyond their data and capacity.

By the nature of their expertise, toxicologists are also concerned with a second issue: the (missed) possibility of protection, rather than simply its absence. On the one hand, casting exposure as the product of constitutive global inequality rightly warns against the naivety of simple solutions (regulation, education, even detection) in protecting Africans from toxic risk. On the other hand, however, this analysis tends to naturalize the absence of protection, dismissing more specific questions about the nature and management of toxic risk. At the opposite end of the spectrum are the Blacksmith Institute's "capacity-bridging" measures of protection for "the poisoned poor": highly effective, even lifesaving, these measures are also necessarily limited in their spatial and temporal reach.[46] Toxicologists aspire to more expansive scales of protective action. Most (if not all) toxicologists in Africa (both locally affiliated and foreign collaborators) have been state employed, and in their research and institution-building efforts they imagine and partially enact more continuous forms of surveillance and control, usually across regional or national territories. In other words, they affirm, even if only indirectly, the possibility of—and the legitimacy of claims to—a protective biopolitics of poison in Africa.

What is on African toxicologists' maps of toxic concerns, and what lies at the edges of and beyond its points of contamination? Over the past decade or two, data on the presence of three categories of toxicants—heavy metals, pesticides, and aflatoxins—in Africa has grown, albeit slowly. Studies have alerted to the risks of heavy-metal exposure associated with, for example:[47] leaded gasoline;[48] oil production and refining (especially in Nigeria);[49] poor disposal, recycling, and burning of waste, including batteries and discarded electronics (e-waste);[50] artisanal and small-scale gold mining (ASGM, which uses mercury and releases lead);[51] industrial mining (especially in South Africa and the Copperbelt);[52] as well as the consumption of contaminated vegetables, fish and seafood, traditional medicines, and cosmetics (notably skin-lighteners containing mercury).[53] As for pesticides, studies have investigated the presence of pesticides (particularly those classified as persistent organic pollutants [POPs] banned since 2004) in breast milk and plant leaves,[54] measured residues on vegetables and in river and drinking water,[55] and examined presumed cases of acute pesticide poisoning.[56] Aflatoxins have been measured in corn/maize, peanut, and cassava products in several African countries, but until very recently both food contamination and human exposure data were very scarce.[57] Data on accidental and voluntary acute poisonings have been compiled from hospital or clinical records, showing the risks posed by pesticides, pharmaceuticals, and, especially for children, paraffin and household cleaning products.[58] With some variations (e.g., in forms and levels of industrial development), this general picture applies to many African countries, including Senegal.

Toxicologists' work multiplies the points of interception of toxic molecules as they are propagated and released by local economic and domestic activity. As with analyses of "dumping," they point to poverty and lack of regulation as major problems: risky, informal occupations such as ASGM as well as ULAB and e-waste recycling occur at the nexus of the dearth of alternative sources of income, the high price offered for metals by foreign buyers, and the lack of oversight over the disposal of batteries and electronics, or over the sale of mercury. Similarly, economic pressures and uncertainty in agricultural production amplify pesticide risks, as do contraband networks, illiteracy and lack of farmer education, as well as the unaffordability of protective equipment and the reuse of scarce plastic containers to store scarce water.[59] Industrial mining and agriculture, powered by foreign capital, are also implicated in environmental contamination; cobalt has been found to travel through the human food chain of the Katanga mining area in the Congo,[60] while there is emerging concern about the presence of pesticides in riverways around irrigated agricultural projects in West Africa.[61] Yet social and economic disadvantage seems to interact in complex ways with access to land, employment, and income in facilitating exposure. In Senegal, for example, toxicologists told me that residents of Ngagne Diaw, as well as communities living around an open-air landfill near Dakar (where they picked and recycled waste) and in villages affected by clusters of (presumed) acute pesticide poisoning, were reluctant to identify toxic exposure and in some cases invoked supernatural causes. Some of the scientists interpreted this as fear of losing homes and livelihoods combined with mistrust of state agents.

Some of these risky substances and activities have attracted international attention. United Nations (UN) agencies, especially the United Nations Environmental Program (UNEP, created in 1972), the WHO, and the Food and Agriculture Organization (FAO), have been concerned with both intergovernmental action and national infrastructures for managing chemical/toxic hazards in the Global South. In addition to coordinating the Basel, Rotterdam, and Stockholm Conventions, UNEP has participated, with the ILO (International Labour Organization) and the WHO, in the International Programme on Chemical Safety, established in 1980, and, with these and other UN agencies, joined by the World Bank and the OECD, the Inter-Organization Programme for the Sound Management of Chemicals (IOMC), created in 1995. Only since the late 1990s have these institutions taken more direct action on the control of toxic hazards in sub-Saharan Africa. These have focused, for example, on fostering national infrastructure for sound chemical management (including poison control centers, in Ghana and Senegal), eliminating stocks of obsolete pesticides, phasing out leaded gasoline, and responding to acute mass poisonings (in Ngagne Diaw and in Zamfara, Nigeria). The Blacksmith Institute, founded in 1999, began working in Africa around 2001 by assisting selected countries in leaded gasoline phase out. Up to 2010, it was involved in twenty or so projects that included decontamination in Ngagne Diaw and Zamfara (as well as in a lead-mining area in Zambia); a few projects, in Guinea, Mozambique, and Senegal, to reduce mercury emissions in ASGM; and various projects to address industrial and urban pollution.[62] While many of these are modestly funded (10,000 to 45,000 USD or so; figures are not given for all projects), some have obtained or leveraged more significant sums from the World Bank. Following the ratification of the Stockholm Convention in 2004, the WHO and other UN agencies initiated a global survey to screen human breast milk for the presence of POPs that included a capacity-building component.[63]

Toxicologists are thus not alone in raising concerns about toxic risks in Africa. Yet few of these international initiatives—besides being fairly recent and modest in their reach—have directly supported the production of data on pathways, levels, and distributions of exposure. Generally addressing known sources of risk, most did not involve toxicological studies (e.g., ASGM projects promoting safer techniques of mercury use without investigating exposure). In Senegal, they have not brought or left much for toxicologists to work with: an action plan for a poison control center (funded through an IOMC-related project, in the hope that the state would support it, which it has done only slowly and partially), a portable blood-lead testing system (provided by Blacksmith to follow up in Ngagne Diaw, it can only function as long as the NGO provides replacements and supplies of testing kits), and some frozen breast milk samples (a junior lab member drew twice the volume needed for the WHO survey and sent half to a lab in Germany for the WHO study and kept the rest, hoping to one day obtain funding for his own study). Earlier, in the 1980s, a UNEP-WHO-FAO project brought an atomic absorption spectrophotometer, which was only briefly used to fulfill its objectives (to measure heavy-metal traces in fish and seafood to monitor marine pollution in West Africa). An exception is Project Locustox, created by the FAO to measure the environmental impact of locust control operations. Based on the argument that pesticide toxicity had to be evaluated where locust control operations took place, and in the ecosystems, climate, and spatial scale that it affected, foreign governments, especially of the Netherlands, joined the Senegalese government in building up ecotoxicological research capacity in Senegal for nearly a decade.

Beyond these intermittent (and one exceptional) provisions of equipment, funding, and research/regulatory objectives, Senegalese toxicologists have asked their own questions, and, with the state paying salaries but giving little toward research, they have mobilized their "own" resources, such as leftover capacity from earlier projects, contacts with foreign scientists, and gifts of free testing or access to lab facilities. What we know about levels of contamination and indicators of exposure in Senegal—and likely in other African settings—owes much to improvisations of capacity that are sometimes productive but often also modest and fragile.[64]

LOSING AFRICAN SCIENCE

Toxicology is certainly not the only science in Africa that has struggled to survive as a publicly funded activity. Nor is the regulation of toxic risk the only state function that has struggled to remain (or become) a source of public protection. The trajectory of toxicology and that of other sciences in Africa follow a broadly shared sequence: from a brief period of growing—but largely promissory—investment in science as an African(ized), national, collective, and development-oriented enterprise (circa 1940s–1970s), followed by a generalized drop in public (both national and international) funding for science in Africa from the 1980s, leading to the stagnation of scientific activity and/or to new "entrepreneurial" strategies for capturing foreign, nongovernmental, or private resources.[65]

There are, of course, variations in this general trajectory, with some sciences in some places and times being the target of more intensive national or transnational investment, especially with the recent rise in transnational funding for global health research.[66] Toxicology, however, has not been a significant or sustained target of attention for the Senegalese state or, as seen earlier, for intergovernmental or nongovernmental organizations. Toxicologists themselves have largely defined their field by deploying and prolonging capacity that was either only briefly or not specifically—and nearly always insufficiently—funded *as* capacity to detect and monitor toxic risks. The state has provided the biggest and steadiest source of money through salaries; indeed, a large proportion of African scientists have been employed by the public sector. In Senegal, however, state salaries seem to have been paid more regularly and at higher levels than in other African countries. Yet national public budgets have rarely paid for laboratory equipment, fieldwork, or other research-related expenses. The main duties of the small number of toxicologists employed by the university have been to train pharmacy students and to take up additional functions (e.g., in education planning, a hospital pharmacy, or the drug control lab). From at least the 1980s, their regular budget could not support research, while their proposal for a national poison control center was put on ice for the next two decades. Research—needed to advance careers and supervise students (the pharmacy degree in Senegal includes a thesis requirement)—came to depend on brief, uncertain sources of support such as international projects, "favors" from sympathetic collaborators, and paid analytical contracts (with the exception, again, of Project Locustox). With the specter of inactivation constantly looming, toxicologists sometimes resourcefully stretched and stitched together remnants of capacity, and sometimes simply waited for the next project or overseas trip. A few gave up on public employment and set up pharmacies or consultancies; others gave up on lab work, investing themselves in teaching or in their additional appointments.

This book, then, tells a familiar story: that of the "abandonment" of public science (and health) in Africa by the state, as it experienced economic crisis from the late 1970s, and, from the 1980s, implemented SAPs (of cuts in state spending and liberalization reforms designed to make African economies more competitive), and, more recently, has been only selectively invested in by newly generous global health donors.[67] This is a story of scientists' experiences of loss, and, for some, of new strategies of survival and success. Loss illuminates change; what was, even if only as possibility or memory, but is no longer. It also illuminates value; what is missed. Following lines of loss can thus help us to understand what scientific capacity, both narrowly and broadly defined, means in settings of (threatened) peripheralization, scarcity, dependence, and stagnation. Capacity is equipment and supplies that were or might have been, the skills to use them, the actions they allowed. But there is also, as Wenzel Geissler has vividly described, the sense of movement and directionality that was activated by functional materials and the qualities of the knowledge it could produce.[68] For the government parasitologists Geissler studied, this was a shared velocity with a collective destination: toward a better future, in which the progressive principle of science fused with individual career ambitions and societal projects of development. As others have pointed out, the sense of movement that animated post-independence African science was also a *synchronous* one, an ability to "keep up" (whether on a distinctive parallel pathway or a converging one) with science elsewhere, that was also underpinned by aspirations to *equivalence* or equity in material capacity and epistemological quality.[69] Such forward-moving, synchronous, equivalent capacity has never been more than a promise. But it is a promise that has grown increasingly elusive, as dependence on projects, collaboration, and contracts has either slowed down or broken up the rhythms of scientific activity and careers.[70] An equivalent toxicology has been associated, in Senegal, not only with innovation or the cutting edge (e.g., the development of new analytical methods) but also with capacity for detection and regulation, that is, the possibility of doing and repeating routine tests to identify and monitor environmental and public health risks. Senegalese toxicology as an active, state-funded science, plugged into functional state mechanisms of food, drug, environmental, and poison control, appeared as a plausible proposition in the 1970s, when the toxicologist Georges Gras proposed a poison control center and set out to measure mercury levels in hair and fish. It is the loss of plausibility of toxicology as equivalent both in its capacity to advance, or keep up, and in its capacity to *protect* that I explore in this book.

Loss of capacity to protect is a central thread in studies of health care and public health in Africa. In their review of the anthropology of structural adjustment and health, for example, James Pfeiffer and Rachel Chapman describe the stripping back of public protections, impeding access to both care and to protective goods such as water, food, and employment.[71] The ethnography of health workers in the public sector has described their reactions, from the 1980s, to a "withdrawal" of the state (manifesting as the drying up of supplies, deteriorating of facilities, and sometimes shrinking or delay of salary payment), and to their own diminished capacity to serve the public. In some cases, workers "abandoned" public service, privatizing care by charging (often illegally) for services and medicines, trading in privileges and favors, or seeking additional or alternative revenue from NGOs.[72] But they have also suffered a sense of moral loss ("demoralization") and sought to improvise care and protection in the face of scarcity, poor working conditions, and the inadequacy or nonaffordability of what they had to offer.[73] Though the term *unprotection* has not, as far as I know, been specifically used in this literature, it is from the sense of loss that it suggests—experienced by those who, "abandoned" by the state, feel they can no longer do their job to care and to protect, as well as by a population that is not or no longer served or protected by a deteriorating public health system—that I define its relevance for describing a tenacious yet largely futile struggle for toxicological capacity. Not in any official dictionary, *unprotection* is defined in a Wiktionary entry as: "removal of protection from something; act of unprotecting."[74] These are exactly the dimensions I seek to underline: a loss of what once was and/or is acknowledged to be possible, and an ongoing, active process that fails to protect, even though it may not aim to expose.

UNPROTECTIVE TOXICOLOGIES

Longings for an African toxicology that once was, might have been, or might yet be equivalently protective raise a question: just how protective would an equivalent toxicology be? Is the better-funded, better-equipped toxicology of better-regulated settings, which Senegalese toxicologists often refer to vaguely as existing out there (France is their main point of comparison), *really* protective? Toxicology is a field defined by a general object—poison and its effects—rather than by its methods or applications; it is a branch of many sciences and disciplines, from chemistry and pharmacology to forensics, occupational health, and environmental sciences. It has many histories, dating back to "the earliest humans," but is generally agreed to have gained in importance and coherence in the twentieth century as a result of two factors. The first is the synthesis of new compounds, from the late nineteenth century and accelerating from the mid-twentieth, and their propagation (as with older poisons such as lead) due to technological progress paired with the intensification of industrial manufacturing, agriculture, extraction, and consumption. The second is the proliferation of regulatory institutions and laws concerned with controlling toxins in foods and drugs and other commodities, in workplaces, and in the environment, which also accelerated after World War II. Though toxicology can be a science of innovation, helping to calibrate novel poisons to kill selectively (pests, parasites, and pathogens but not hosts or bystanders), and a forensic science (cause of death, doping control, etc.), it has come, in the postwar era, to play an important societal role as a science of regulation.[75]

As a science able to call into question the safety of lucrative molecules and of their profitable uses, perhaps even shake the very foundations of industrial society, toxicology was, as Nathalie Jas has suggested, "potentially subversive."[76] Yet as she, Soraya Boudia, and others have concluded, this potential was straitjacketed by a compromise: between protection and production.[77] This compromise, as Christopher Sellers has shown, penetrated toxicology's very methods, which were centered on the mechanistic determination of safe thresholds (on the basis of animal dose-response tests) of toxic concentration, that is, "the dose makes the poison."[78] Developed during a time when research on toxic hazards was largely funded by industry (in its own laboratories or via university departments),[79] this threshold-based toxicology was reassuring: exposure was measurable and therefore could be controlled.[80] After World War II, toxicology's methods were enshrined in new or expanding national regulatory institutions (in the United States [where toxicology has generated the most social scholarship], the principal ones are the Food and Drug Administration [FDA], the Occupational Health and Safety Administration [OSHA], and the Environmental Protection Agency [EPA]), as well as international standards for food, drug, environmental, and occupational safety.[81] These methods were slow to catch up with the new risks posed by the exponential increase in new chemicals and their pervasive presence in open, dynamic environments, lending, as Michelle Murphy puts it, "a narrow shape to what counted as a significant chemical exposure" and thus producing a "domain of imperceptibility."[82] Toxicology's methods were unable to define or control the toxicity of postwar pesticide-saturated agricultural landscapes,[83] of the late twentieth-century synthetic office building,[84] or of the endocrine-disrupting effects of low-level exposures to plastics.[85] Toxic ignorance is also produced by what goes untested: the EPA and FDA, for example, have been criticized for testing only a tiny fraction of chemicals in circulation and of produce for pesticide residue monitoring, and for inadequately sampling in hazard assessments.[86] If science has also, at times, been clearly and scandalously manipulated by industrial interests,[87] its "powerlessness" to generate protective knowledge is, for Boudia and Jas, "systemic," built into "the very functioning of [regulatory] systems." They conclude, "Despite the immensity of the activity they have generated, these systems have not allowed for the production and accumulation of real knowledge on toxic substances."[88]

My aim here is not to set up a detailed comparison between Senegalese toxicology and its better-equipped counterparts. It is to ask the question: If toxicology everywhere is unprotective, then what, if anything, is distinctive about toxicology in Senegal? To some extent, "African" toxicology is simply an extreme point on a spectrum of unprotectiveness in the exposure sciences, that is, one instance of a science that has generally failed to keep up with the proliferation and complexity of toxic risk (an "archaic" science, as Murphy has said of mainstream, regulatory toxicology).[89] Yet the protection/production compromise that has held toxicology *back* has also been held *up*, in wealthier economies, by a degree of minimal protection. Even an "archaic" toxicology can, for example, identify groups of children who are at risk of lead poisoning, measure pesticide residues in food, or detect the enzyme-inhibiting effects of exposure to some types of pesticides in blood. In Senegal, the capacity to perform and repeat even such basic tests, using standard analytical methods, in order to detect substances and measure concentrations that can then be compared to accepted safety standards, has often been missing or partial. It has never been taken for granted. While toxicologists have, at times, managed to do this on tiny scales (thus pointing, for example, to the contamination of artisanal peanut oil and paste with aflatoxin, or of tomatoes and citrus by heptachlorine residues),[90] their calls for "regular monitoring" have marked off a wide expanse of missing knowledge—the results of repeated tests that neither they nor another laboratory were likely to perform—from the fractional coverage of their improvised capacity. Their dependence on external support (and the limits this has imposed on their capacity), the "fictional" nature of the "regular monitoring" for which they often call (yet which gives meaning to their small-scale work), and their sense of failing to measure up not only to the nature of toxic risk in Senegal but also to a better, more equipped toxicology in the Global North: all this adds a specific (postcolonial) quality to the quantitative "end of the spectrum" position of Senegalese toxicology.

Whether varying in degree or in kind, the unprotectiveness of toxicology has not been accepted passively by all of its practitioners. As Kim and Mike Fortun, Scott Frickel, and Michelle Murphy have shown, some groups have defended toxicology's professional ethos as a public-service science, or as a "civic science,"[91] by protesting and pushing against the limits placed on their ability to know and to protect. The Fortuns describe a "sense of the civic" among American toxicologists that is anchored in a narrative of postwar regulatory expansion that underpins toxicologists' "commitment to practical knowledge."[92] In the pursuit of this ethos of regulatory application, limits to knowledge, or "not-knowing," provokes, they observe, ethical anxiety.[93] Historically, the kind of not-knowing toxicologists have worried about is the biased or misguided manipulation of their methods or results. Murphy provides a good example. In the 1980s, a large group of EPA employees formed a union, Local 2050, to claim their right to a "neutral" workplace against the threat—to the institution's founding ethos of state protection—of the pro-industry, antiregulation influences allowed in by the Reagan administration.[94]

The fight against not-knowing has also taken the route of promoting methodological innovation. From the late 1960s, scientists engaged in what Frickel has described as a "modest . . . reform movement" to institutionalize genetic toxicology (the study of toxic effects, notably of mutation, in genes). These "scientist-activists," as he calls them, sought to overcome the not-knowing imposed by old disciplinary boundaries and regulatory structures, thus laying the foundations for a "new public-service genetics."[95] In the case of toxicogenomics, the study of responses of the entire genome to toxic exposure, too much information can also, as the Fortuns observe, become a form of not-knowing, in the sense that complexity and uncertainty can obscure pathways to regulatory applications. They describe how "caring for data" in order to inform regulatory practice even in the absence of certainty was thus imbued with ethical significance: "a sense of the civic that depends on and mandates information infrastructure."[96]

I would not go as far as to call Senegalese toxicologists *activists*, even in the modest sense proposed by Frickel, for they could have taken a much more public role in diagnosing and denouncing the conditions that generate exposure (e.g., by working more closely with the Pesticides Action Network) as well as the inaction and inefficacy of the state, or in putting poisoning on national and global health agendas (as Blacksmith has been doing recently, and, to some extent, the staff of the CAP). Yet in their very struggle to survive and succeed as scientists, and to create or maintain the three toxicological institutions that I studied, they have improvised and imagined a more capacious and protective toxicology, thereby "refusing" the forms of not-knowing that threatened the "civicness" of their practice (as in the cases described previously). Improvising with scarce resources, as Julie Livingston and Claire Wendland have suggested for nurses and medical students, can activate a collective ethos of care and responsibility, one that has, historically, been defined as civic and national. To some extent, then, improvisation stretches the limits of capacity toward imaginations of good medicine and nursing that are defined not only by technical but also by moral criteria of efficacy and value. Still, if capacity stretching can be seen as a form of protest against the unprotectiveness of poorly equipped medicine or science, it can lead not only to moral imaginations of responsibility and commitment but also to moral illusions of protection, tied, for example, to regulatory futures (of "regular monitoring") that may never come. The challenge, then, is to discern, in improvisation, a will to protect while recognizing its fragilities and futilities.

STUDYING THE RHYTHMS OF CAPACITY

Lost capacity, improvised capacity, missing and future capacity; these, my informants (people who have worked in the three sites I studied) differentiated, as Geissler's parasitologists did, in terms of temporal qualities of rhythm and direction. They referred to past times when scientific activity moved faster, kept pace, and filled up time, when it could be synchronous, continuous, and cumulative. They described the slowing, intermittence, and waiting that resulted from sporadic and uncertain funding, broken-down equipment, and trips overseas. At the poison control center, hoped-for futures were of durable routine surveillance and response, while at Project Locustox, the prolonged time of cumulative ecological observation was argued to be crucial for building an ecotoxicological regulatory infrastructure.

This book is cadenced by these rhythms. It seeks to decipher their meanings as expressions of value, that is, of the goods associated with active, well-equipped science; of the material but also moral threats of inactivity/inactivation; and of how scientists would like to see themselves as committed public scientists, as resourceful African scientists, and as equivalent global scientists. In other words, I follow the contours of scientists' own narration of their history and explore how its rhythmic qualities (sometimes explicit, sometimes as I interpret them) define capacity and its corollaries: scientific virtue, the advancement of knowledge and of careers, public service and protection. Thus, the "rhythms of capacity" delineate the loss and pursuit of "good" science in times of shifting hopes and uncertain, often scarce resources.

This temporal vocabulary picks up on aforementioned work on memories and histories of postcolonial African science. It also takes inspiration from a broader set of anthropological studies that have described experiences of material decline and reversal in Africa in terms of shrunken, interrupted, fragmented, and "leaping" temporal horizons: what Jane Guyer has called "the temporality of lived economies."[97] In a landmark 1995 article, Janet Roitman and Achille Mbembe vividly describe the nostalgia, incomprehension, and uncertainty with which Cameroonians responded to a sudden 50 percent currency devaluation (in 1994, as a measure of structural adjustment aiming to reduce the price of exports). The economy, which had only recently seemed to be "on a continuous and irreversible path of progress," now manifested as unfinished buildings, decaying urban infrastructure, and unpredictable, shrinking salary payments.[98] In the Zambian Copperbelt, James Ferguson contrasts "expectations of modernity" of the post-independence decades—the progressive temporalities of economic and political emergence that promised a "joining up with the world"—with, in the 1990s, a feeling of loss and of "abjection," of being "thrown aside, expelled" from membership in global society.[99] Economic decline and reform also weakened the hold of African states over collective experiences of time. Shared memories, histories, and anticipated futures were central to the ambitions of African postcolonial nation-building. These were orchestrated through economic planning, commemoration, monuments, public employment and services, patronage of the arts, and infrastructural investments.[100] From the 1980s, the actions of the post-developmental state became increasingly episodic, delayed, or unpredictable, while the "event-based" presence and narratives of the churches and NGOs took over many of its functions. In Togo, writes Charles Piot in 2010, "the linear time of the dictatorship . . . and the continuous time of the ancestors is being replaced by a noncontinuous temporality, one that is 'punctuated' (Guyer, 2007) and event-driven, and one that anticipates a future while closing its eyes to the past."[101] Others have described new temporalities of "projectification" generated by NGO and transnational interventions, or the "syncopation" of day-to-day survival.[102] Science, with its heavy reliance on public funding and its strong associations with development and progress,[103] in combination with worldwide trends in the cost and technical sophistication of scientific research, has been particularly vulnerable, in Africa, to such temporal interruption and fragmentation.[104]

At the same time, as I point out, scientists themselves also, in their improvisations as well as in their memories and aspirations, seek to set the tempo of their own work—or at least their narratives and fantasies of this work—thus actively pursuing what Adeline Masquelier has called "meaningful temporalities."[105] In the case of toxicology, a science whose identity is, in Senegal as elsewhere, strongly linked to its regulatory applications, meaningful rhythms are not just those of advancement (of innovation, progress, development, success, etc.). They are also those of continuous activity, of the regularly repeated analyses of surveys and monitoring, and of incremental accumulations of data. The value of such routine and regulatory rhythms of scientific activity has received little attention in African, or other, settings, especially other than as a form of "time-discipline" (as in E. P. Thompson's seminal essay on the inculcation of new modes of industrial production)[106] or beyond Weberian associations of routinization with rationalization, bleakness, and disenchantment.[107] Where continuous and incremental time is fragile, where work and lives are constantly disrupted, the promise of routine temporalities might be invested with both practical and moral value as a source of order, security, and protection.

Deploying and attending to "rhythmic" descriptions of succession, intersection, and discrepancy between pasts and presents, this book is both historical and ethnographic. The majority of its sources are historical in that they are "from" the past: laboratory spaces and equipment that have remained; documents that have been archived, left, or put away; and stories told about what was (only the last chapter is based more substantially on ethnographic observation). In addition, my narration is historical, in that I refer to the past as being *in* the past, which I order chronologically and periodize. I describe change and package sequences into times of greater or lesser resources, capacity, and activity. And yet, this book is mostly about the present. Relying to a large extent on oral history to guide my overarching narrative, and to interpret a very fragmentary textual and material record of past activity, I listened to stories and looked at objects as they existed in the present.[108] I made an effort to get to know this present; I paid attention to the sites where I found documents (most were not formally archived but kept or left in situ) and where I conducted interviews, and spent additional time interacting with these spaces and their occupants informally (though more so at the poison control center and the university laboratory than at "Locustox"). It was through my encounters with people, spaces, documents, and equipment in this present that I glimpsed what lost, gained, elusive, illusory, and hoped-for capacity might mean to toxicologists (then but especially now). This present was experienced by my informants and materialized in laboratories or their absence, as one of incapacity, or, more specifically, of diminished (no longer) and anticipated (not yet) capacity. I do not fully share the sense of nostalgia and optimism they often projected and instead underscore past constraints on their capacity to detect and to protect as well as its future uncertainty. Yet I also take seriously memories and hopes of "better times" as indices of "better toxicology." If this book is also about toxicology in the past (I do attend to sources as *both* records and remains), it is mostly about the meaning this past acquires from and gives to the present, during the time I spent studying it, over a period of about eight months between January 2010 and March 2011.

TEMPOS: STRUCTURE OF THE BOOK

The chapters in this book are ordered chronologically and framed according to the tempos that my informants remembered or described for distinct periods. An exception to this structure is chapter 1, which is polyrhythmic and enters the laboratory of toxicology and analytical chemistry at the UCAD in Dakar during a present time of inactivity (at the time of my fieldwork in 2010). In this chapter, I describe the challenges and reflect on the stakes—for defining how capacity is made and kept—of recovering the rhythms that once animated its equipment. While primarily a reflection on what to make of a stilled but once active material record of scientific activity, this chapter also provides an overview of the lab's history that serves as a map for the periods described in chapters 2, 3, and 5.

Chapter 2 follows the history of the university lab from the early 1960s to the early 1980s. The post-independence decades were the time of *la coopération* (overseas cooperation), that is, of French technical assistance to its former African colonies. French nationals still occupied most senior positions at the university, including in toxicology and analytical chemistry. Yet technical assistance promised the mutual advancement of both expat and African scientific careers, as well as of science itself and of African development. Focusing on the aspirations to *advancement* expressed by different members of the lab—a former colonial pharmacist, a French academic toxicologist, and a Senegalese technician—this chapter illuminates tensions between, and contradictions within, distinctive visions for the Africanized toxicology that might emerge in Dakar.

Chapter 3 examines the pursuit and value of *routine rhythms* of scientific work in the university lab from the early 1980s. At this time, the first Senegalese PhDs in toxicology and analytical chemistry returned from France to replace the French *coopérants* (technical assistants) in the lab.

Cuts in both French assistance and Senegalese state funding threatened to break up the tempos of scientific work in the university laboratory at the time when its leadership was being Africanized (or "Senegalized"). International investments in the lab's equipment provided opportunities to stretch and project analytical capacity toward the regular monitoring of toxicity in Senegalese environments. While exploring the civic value invested in regular rhythms and routine science, this chapter also presents the memory and ideal of protective toxicology during this period as a *fiction*.

Chapter 4 is about the prolongation of a transnational collaboration that aimed initially to evaluate the environmental effects of chemical locust control, and then to develop a "Sahelian" ecotoxicology for assessing pesticide toxicity. Focusing on the arguments put forward for continuing to invest in Sahelian ecotoxicological research, this chapter examines *prolongation* as both an epistemological and political project to link the "Sahelization" of ecotoxicology's methods to its durable relocation in Sahelian institutions. Those who made these arguments understood infrastructure and resultant capacity as an accumulation of connections between ecosystems, institutions, scientists, data, equipment, and methods. The fragile success of the transition from collaborative project to permanent local institution raises questions about the kinds of support, and the kind of scientific work, needed to make an environmental science responsive to national imperatives of regulation and protection.

Chapter 5 is about poison control in the making during a time of renewed state provision. In 2010, I observed the CAP's director and staff seeking to initiate, as soon as possible, regular and continuous rhythms of surveillance and response. This "hasty routinization" was set against the prior temporalities of delay and crisis attending to the center's creation, and sought to evoke new opportunities for an expanded biopolitics of poisoning. Yet as they moved toward bureaucratic routines of government, center staff also distanced their project from the Senegalese state as an uncertain and partial provider while working out how to complete the construction—both literal and metaphorical—of their institution.

1 · After Interruption: Recovering Movement in the Polyrhythmic Laboratory

The distinct aesthetics of decades past jostle on the lab's surfaces. The oldest layer is of chipped tiled teaching benches, rusty gas taps, and heaped, dusty glassware. Professor Fall designates these with a sweeping gesture: "All antiques!"[1] Large gas canisters, also flecked with rust, bottles of chemicals that have expired more recently, and the square heft of a broken-down gas chromatograph (GC) bear witness to less distant, but now ended, analytical activity. Rose Diene, a longtime administrator in the lab, calls these "wreckage."[2] The newest layer of materials includes secondhand equipment donated by a French lab and two kinds of portable testing kits, one designed to monitor drug quality and the other to measure blood-lead levels. These are, in theory, in working order but await supplies, set up, and maintenance not covered by the lab's budget. The portable kits are set in motion at intervals by chemicals, replacement parts, and field costs provided by foreign funding organizations for monitoring projects, while a donated HPLC (high-performance liquid chromatography) system is used occasionally by means of a small cache of reagents picked up on overseas visits. Meanwhile, freezers keep urine, breast milk, and river water suspended in time until they can be carried overseas for tests of traces and markers of pesticides and heavy metals. When I spent time in the lab in 2010–11, I saw barely any activity at the bench. Instead, I heard complaints that "nothing work[ed]." I was told that, due to the lack of chemicals, practical lab exercises had to be canceled or demonstrated by the technicians (i.e. not performed by students) and that outside requests for analytical services were turned down. Research projects and advanced degrees could only be completed overseas.

In 2010, these stilled remains of past scientific activity are the material manifestation of toxicology (and analytical chemistry) at Dakar's national public university, the Université Cheikh Anta Diop (UCAD). This lab was the first and main, and at times the only, site where toxicology was practiced and taught as an analytical and experimental science in Senegal. As a division of the Faculty of Pharmacy, it has been primarily a teaching and administrative unit. It was created as part of the post-independence extension of pharmacy education from a technical degree to the full state qualification, on par with a French degree. A toxicology section was first mentioned in the faculty's bulletins (*Bulletins et mémoires de la faculté nationale de médecine et de pharmacie de Dakar*) in 1961, and the first state degrees in pharmacy were conferred in 1962.[3] Administrative designations varied before 1970, and the unit split into two subdisciplinary units in 2007 (one for toxicology and hydrology, and the other for analytical chemistry and bromatology). Yet for most of its history, the lab was known as the Laboratoire de chimie analytique et toxicologie (Laboratory of Analytical Chemistry and Toxicology, or LCAT). I refer to "the lab" in the singular under its older and best-known administrative-disciplinary heading. This simplifies things but also conveys the continued sense of shared history and identity among both labs' current members, as well as their complementary interests in detecting contaminants.

Pharmacy students are taught theory by the lab's faculty, and, in their fourth and fifth years, perform practical exercises in its two purpose-built teaching labs. They can also choose a lab member as their supervisor for the thesis work they must complete and defend in their final year of study, as a graduation requirement. For these thesis-year students, and for lab staff members, who must publish to be promoted and who are occasionally enlisted in collaborative projects, this is also a research lab. There are designated rooms for research, where most of the analytical equipment is kept. Other rooms serve as offices. Research activity has at times been supplemented by (and sometimes mimicked) routine testing for toxic traces and indices of contamination requested by private clients and governmental agencies. In 2010, however, such requests were being turned down unless clients were willing to pay for shipping and analysis elsewhere. In the combination and intersections of student and staff research, analytical services, and international collaborative projects, the lab has performed, though usually only partially, a broad range of functions: from the provision of medicolegal expertise to tests of food quality and suspicious substances and the monitoring of indicators of exposure and contamination. In addition, senior staff have been cross-appointed to ministerial functions, in particular at the hospital pharmacy of a neighboring teaching hospital campus and at the national drug control laboratory (Laboratoire National de Contrôle des Médicaments, or LNCM), and have initiated at least two attempts, the most recent one successful, to create a national poison control center (Centre Anti-Poison, CAP).

This lab, then, is a key site from which to trace the formation and fate of toxicology as a Senegalese science, and to examine changing capacities to detect, define, and regulate toxins as posing particular kinds of problems in Senegal. What can the lab's current materiality, sedimented over time and now largely deactivated, reveal of this history? What pasts and processes, perhaps also futures, can be salvaged from these "antiques" and "wreckage"? In this chapter, I explore ways of recovering past and potential movement from the lab's inactive materials. I seek, in particular, to decipher the rhythms and directions that once animated its rooms and equipment, and to explore how this movement sustains and haunts scientific aspirations over time. I am not interested in making matter speak *for*, or in the absence of, people. While much of this chapter focuses on apparatus, reagents, gas taps, benches, signs, and reports, and considers how these carry and convey traces of former analytical activity and future horizons into the present, my aim is to reflect on *who* makes and moves capacity, and for whom.[4] I am also guided in large part by the memories and recent experiences of those who told me about the lab's past. Their stories gravitated around the (non-)functionality of materials. Critical to performing and narrating capacity as functional in the past, now-deactivated equipment make palpable—for both the historian and the scientists who inhabit the lab and have depended on it to sustain their careers—the difficulties of making knowledge under chronic resource constraints. And beyond this: the provision, design, and maintenance of the material conditions of science also enact power, generosity, autonomy, dependence, abandonment, and so on. In other words, these things were and are about relationships and investments between people.

In developing three possible readings of the lab's material record—as things remaining, as things designed, and as things kept—I set up a discussion of what these imply for how we describe the (re)making of scientific capacity. The different actions by which things have made their way into the lab's future bear differential imprints of the actors who invested in the lab's capacity: foreign funders and collaborators, the Senegalese state, and lab members, both senior and subordinate. My methodological reflection on how to recover movement also initiates a broader reflection on the meaning of these recovered movements, which coalesces around these three questions: What directions and futures—of scientific advancement, public health, personal careers, institutional histories—have project endings, dwindling budgets, and equipment breakdown interrupted for those who have worked in the lab? How have toxicologists, whether as students, assistants, faculty members, or civil servants, pushed back against material interruption to remember, imagine, and enact continuous, regular, and/or progressive rhythms of scientific activity? And what constraints, material and imaginative, are ultimately revealed by the failures of these efforts to push back—this "will" to work, to know, and to protect—in procuring protections and betterment for the Senegalese people? Finally, this chapter sketches out an overarching periodization of the lab's history that provides chronological orientation to the two chapters that follow.

THE MEMORY OF REMAINING THINGS

Enter the lab. Upstairs, through a door, is the toxicology teaching lab. Along the wall is a metal filing cabinet; on it, differently shaped and tinted bottles of chemicals are lined up. The expiry dates on their labels range from the 1970s to the early 2000s. On the shelves' edges, masking-tape labels announce the topics of practical lab exercises: "barbiturates," "cannabis," and "salicylates." Downstairs, in the analytical chemistry teaching lab, other substances are scrawled in chalk at the end of built-in benches: half are "milk," the other half "flour." The rows of tiled benches are identical in both labs; the vents and the gas and some water taps are in obvious disuse; one has a warning not to be opened. In a box at the bottom of a cabinet, glassware is piled up and dusty, but upstairs, on one visit, beakers, flasks, and test tubes have been cleaned and attractively, but randomly, displayed along the top of the bench dividers.

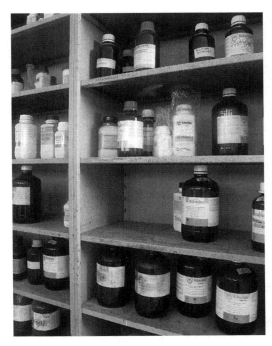

FIGURE 1.1. Chemicals on shelves, toxicology teaching laboratory. Photo by Noémi Tousignant, Dakar, 2010.

FIGURE 1.2. "Don't use the taps," analytical chemistry teaching laboratory. Photo by Noémi Tousignant, Dakar, 2010.

Downstairs, off the side of the teaching lab is the analytical chemistry research lab. Here, on the counter, sits the square heft of a two-decade-old gas chromatograph, and on the floor is a pile of suitcase-sized black plastic boxes that contain the components of a simple portable lab to test drug quality. The toxicology research lab is down the corridor, among the offices. The tiled benches and relatively new equipment are kept spotless by Madame Bassene's daily ministrations, yet this room is no more filled with action than those filled with "antiques" and "wreckage."

My first questions were about the origins and arrivals of these materials. These elicited stories about various moments of transnational collaboration. The oldest equipment was associated with la coopération, the French technical assistance provided, after political independence in 1960, under agreements between the governments of France and Senegal in various areas such as defense, culture, and higher education. For the few who were in the lab then, this was a time, which lasted until the early 1980s, of regularly renewed supplies; a time when, Rose Diene now recalls with wonder, "we ordered things from France . . . and they came!"[5] These things, according to Babacar Niane, were "not big apparatus," not like the automated analytical machines that later "projects" would bring. Rather they were lightweight materials to be manipulated by hand: flasks, burners, pipettes, and so forth.[6] This may not be entirely true. Georges Gras, the head of the lab throughout the 1970s, had acquired a flameless atomic absorption spectrophotometer (AAS) to analyze traces of mercury.[7] Still, published reports from that time emphasized the need, at such a distance from suppliers of equipment, parts, chemicals, and maintenance expertise, to develop inventive, low-tech manual methods of analysis for detecting and dosing toxic traces.[8] The continued use, over decades, of a set of practical teaching exercises, on unrenovated benches strewn with old glassware, strengthens this identification of these objects Professor Fall calls "antiques" as a legacy of the time of Franco-Senegalese cooperation.[9]

The "wreckage" of machines forms the remains of another era: the era of "projects." This era began, lab members agree, just after Gras's departure circa 1982 and the arrival of freshly graduated Senegalese faculty members from France (though, again, documents suggest a less definite demarcation, for there were "projects" in Gras's time too). From a joint UN agency project that began in 1983 to a vaguely defined "Italian" project in the early 1990s, three international collaborations, all concerned with creating capacity to monitor food and environmental contaminants, provided three fairly sophisticated systems of automated analysis: the machines. Identified by lab members as "project machines" and later also as "wreckage," these evoke an image of waves washing up heavy pieces of lab equipment and filling the lab with activity, then receding, taking with them the supply of "consumable" supplies (e.g., reagents and reference substances), maintenance, and expertise.

The still-functional portable testing kits are also from foreign sponsors acting within the objectives of specific projects, but they are certainly not machines. Both trademarked, the Minilab and LeadCare II are designed to be lightweight and easy to use, even without a lab infrastructure; that is, they provide capacity to perform a single or narrow range of tests without requiring a broader investment of people, materials, and institutions. The Minilabs in Dakar were identified for me as coming from "USP" (U.S. Pharmacopeia) for a sentinel-site drug-quality monitoring project begun in 2002, while the LeadCare II was provided by the U.S.-based NGO Blacksmith Institute for follow-up monitoring after the mass lead poisoning in Ngagne Diaw (see introduction). In 2010, both types of kit are still being periodically supplied with chemicals, replacement parts, instructions, and supervision by the American organizations. Stacks of cardboard boxes of reference substances have been shipped from the United States (only for those classes of drugs targeted by the USP program). A replacement Lead-Care II has arrived, provided by Blacksmith, but only after months of delayed blood testing. Continued activation depends on continued support from overseas. Furthermore, the trade-off for "lightness" is a limited analytical scope. Minilab components provide only for screening tests of the conformity and purity of medicines, while the LeadCare II has a ceiling of 600 ppm of lead in blood. A test I watched on blood from a baby, born *after* epidemic poisoning was averted, now suffering from encephalitis, confirmed only that its lead concentration was higher than this. Such blood samples, as well as drugs that are identified as substandard by Minilab testing, have to be sent elsewhere (a French lab in the first case, the LNCM in the second) for more accurate results. Although they are still active, the testing kits manifest the limits and vulnerability of the capacity flows of which they are part, and they, too, could soon become lifeless remains.

Finally, I was shown some gleaming apparatus in the toxicology research lab room, but then quickly told that most of this secondhand equipment, which was handed down by a French "partner," did not yet work. Donated by an environmental toxicology lab in Dunkerque, France (I also

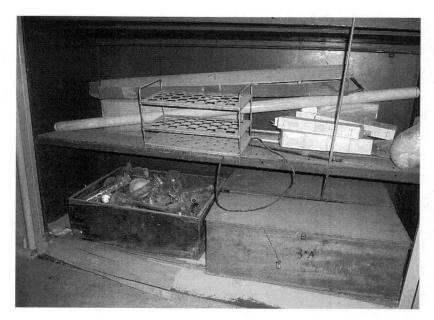

FIGURE 1.3. Dust-covered glassware, analytical chemistry teaching laboratory. Photo by Noémi Tousignant, Dakar, 2010.

FIGURE 1.4. Gas chromatograph, analytical chemistry research laboratory. Photo by Noémi Tousignant, Dakar, 2010.

heard of an air sampler donated by a lab in Rouen), this equipment is one material manifestation of long-standing exchanges between French and Senegalese scientists. Although the Dakar-Dunkerque collaboration was formalized by an agreement in 2004, it stems from a longer history of personal and friendly exchanges between Dakar lab members and French scientists that emerged in the aftermath of la coopération. Doudou Ba, who became head of the lab in the early 1980s and recently retired, told me that one departing French technical assistant maintained his relations with the lab because he "understood our difficulties."[10] The main benefit of these relations has been access to better-equipped French labs for Senegalese faculty during their doctoral research or short-term "study visits," and where analyses on samples brought from Senegal have been performed for free. Thus most of the flow has been toward France to palliate the lack of equipment in Dakar. This secondhand apparatus is a rare instance of travel in the other direction, yet the activation of this equipment still depends on the continuation of friendly relations and favors. Only the HPLC, I was told in 2010, gets occasional use because it was set up by a visiting lecturer from Dunkerque and is supplied with small amounts of reagents brought back by a junior faculty member who is doing her PhD in Dunkerque, where, she says: "I ask for a bit of this, a bit of that."[11]

Thus "antiques" and "wreckage," portable kits and hand-me-downs can be seen as that which *remains* in Dakar from successive (in some cases overlapping) phases of foreign investment and abandonment. Most are now useless or only sporadically useful; they underscore lab members' lack of autonomous ability to replace, repair, or reactivate them. They arrive and are set in motion by provisions made by "Northern" or international institutions within the scope of specific objectives and time frames, as in the case of projects, or given as "gifts"—rather than entitlements or benefits earned—of goodwill, cooperation, and friendship. African labs are not the only labs dependent on fluctuating external sources of funding and equipment. Yet labs in the Global North not only have access to larger pools of material resources but also to more opportunities for obtaining these by enunciating specific research objectives (even if these may be shaped by funder priorities). Very little funding or equipment for research in the toxicology and analytical chemistry lab in Dakar seems to have been directly tied to the articulation of a research project or program by members of the lab, except by French nationals during the time of Franco-Senegalese cooperation. Thus layers of equipment appear as "leftovers" of unpredictable

FIGURE 1.5. LeadCare II (*bottom-left*) and donated equipment (*background*), toxicology research laboratory. Photo by Noémi Tousignant, Dakar, 2010.

transfers of analytical capacity originating in the Global North and, in most cases, of which the uses and utility are also defined elsewhere. This is not the only way to read the history of the lab's capacity. But before moving on, what does this reading of things as remains say about the logics and effects of successive waves of transnational exchange?

In the years after Senegal's independence, analytical capacity at the University of Dakar was fostered as part of a broader commitment to maintaining "equivalent quality" with French higher education. Equivalent quality was explicitly pledged under the Franco-Senegalese agreements regarding the University of Dakar in 1961.[12] Founded in 1957 as a French university (a *metropolitan* rather than a *colonial* institution, as Thomas Eisemon and Jamil Salmi point out, sometimes called the "eighteenth university of France"),[13] the University of Dakar, even after it was declared a "Senegalese institution" in 1961, remained closely tied to French national education. For a time, it conferred degrees that were registered by both the French and Senegalese ministries of education.[14] The university's "often cited official sub-title" was "a French university designed to serve Africa," but throughout the 1960s, about a third of its students were French.[15] The

French government continued to pay a portion of its budget, including the salaries of French technical assistants. The Senegalese government continued to request large numbers of such coopérants, and the university's curriculum was modeled on the French system.[16] In May 1968, students in Dakar protested the lack of curriculum reform and the slowness of Africanization of positions of authority (and more generally the maintenance of neocolonial Franco-African relations). France reduced (but did not end) its financial contributions and administrative oversight, and curricular reforms were initiated. Degrees from Dakar were no longer automatically valid in France, but those in medicine and pharmacy were still reciprocally recognized. French commitment to the quality of Senegalese higher education was reformulated in terms of achieving an "international standard."[17] Many departments and units across the university continued to be headed by coopérants into the 1980s, including the toxicology lab.

Equivalent quality meant that African students gained new pathways of professional mobility. It also, however, allowed children of the many French expats in Senegal in the 1960s to begin their degrees in Dakar and easily transfer to French universities. Coopérants could also easily move up, via promotions they obtained in Dakar, to good positions in France. Thus Diene and Niane's memories of continuous flows of lab equipment as specific to the time of la coopération suggest that analytical apparatus ordered by French nationals—which they and their Senegalese and French colleagues, students, and technicians manipulated to obtain practical skills, degrees, and promotions—generated capacity for mutual benefit and advancement.

While Franco-Senegalese cooperation was a university-, even nationwide project to nurture national technoscientific expertise, the projects that deposited machines (the AAS, GC, and HPLC) in the lab aimed for more targeted analytical and regulatory capacity. Lab members remember three projects in particular: WACAF/2 (1984–1989), Project Locustox (1989), and "The Italian Project" (early 1990s). On the basis of uneven evidence (the second is well documented, the first less so, while, for the third, I found only imprecise and contradictory recollections), these can be characterized as intergovernmental efforts to strengthen national capacity at a regional level to monitor the contamination of food and environments with pesticides and heavy metals. They entailed standardized techniques, not innovative ones, using fairly heavy pieces of equipment, which were provided along with training (through internships in Europe) and guidelines, and sometimes with complementary supplies (spare parts, glassware, chemicals, apparatus for preparing samples). I give more detail on these projects in chapters 3 and 4; what is important here is that, in contrast with the broad aspirations to equivalence that are carried by remains of la coopération, the machines appear as remains of past investments that aimed at jump-starting long-term regulatory monitoring of environmental and health risks (marine pollution, the disruption of ecosystems by locust control operations, possibly pesticide residues in food) that would then be carried on by the Senegalese state.

While the "antiques" and "wreckage" were brought in to *build* durable capacity (for education or for regulation), the Minilab and LeadCare II appear, in contrast, as capacity-*bridging*. That is, they make it possible to postpone or evade broader problems of missing capacity to fulfill an immediate need: to identify problems of drug quality and to monitor pre- and post-intervention blood-lead levels in Ngagne Diaw. The online pamphlet for the Minilab describes it as "a complete laboratory in two suitcases" that can "*bridge* the capacity gap in regular drug quality monitoring on a national level in low-income countries."[18] In Senegal, a set of Minilabs was deployed in 2002 as part of the Medicine Quality Monitoring project (MQM), overseen by the USP Drug Quality and Information program (DQI) and funded by the United States Agency for International Development (USAID).[19] Kept in the university lab, the suitcases were taken out three times a year (then six, from 2008) to five sentinel sites across the country.[20] The DQI does invest in more continuous and durable capacity for "drug quality assurance and control" through the Senegalese pharmacy regulatory authority and the LNCM.[21] Its material legacy in the university lab, however, has a limited analytical scope, useful mainly as a screening step that requires the backup of the better-equipped LNCM. Perhaps more importantly, this capacity is circumscribed by the disease-control priorities set by global health initiatives. MQM initially tested only antimalarials, expanding to antiretroviral and tuberculosis drugs in 2006: additional support for the program was provided through the U.S. President's Malaria Initiative (PMI), created in 2005.

The analytical capacity brought by the LeadCare II is even more limited in its scope and duration. The website of the testing system promotes its ability to provide instant results anywhere, as a means of identifying children at risk of lead exposure despite the difficulties of access to health care in the United States.[22] Blacksmith's provision of the apparatus to the CAP team (which keeps it in the toxicology lab because it does not yet have its own lab facilities) places the testing kit on a global map of "toxic hotspots" such as Ngagne Diaw. Because the LeadCare II is not designed to test for higher concentrations or other contaminants, and because of dependence on Blacksmith to replace and supply the apparatus with very specific (and expensive) kits of consumables (reagents, droppers, capillary tubes, etc.), Senegalese toxicologists have very limited flexibility and autonomy in their use of this testing capacity. Two lab members, for example, complained that it was not very useful for their studies of exposure to heavy metals in a community living on or around an open-air landfill, which, incidentally, is where Ngagne Diaw's contaminated sand was first moved.[23]

While the older, heavier "machines" were meant to create a fully functioning lab to serve (inter-)governmental regulation on a national or regional scale, the Minilab and LeadCare II's lighter capacities trace out cartographies of drug quality and toxic contamination as problems of global rather than national public health.[24] They connect to "global epidemics" (of poor-quality drugs, of life-threatening pollution) rather than with contiguous territories of potential contamination.[25] Neither is provided as isolated capacity; as mentioned previously, the DQI program reinforces other aspects of pharmaceutical regulation, while Blacksmith has supported both short- and longer-term educational activities in Ngagne Diaw (recently, a job skills training program). Yet the mobile apparatus is designed to work even in the absence of functioning (national) laboratory and regulatory infrastructure. Like the humanitarian technologies studied by Peter Redfield, they do not assume existing (state) capacity, and seek to render it dispensable to fulfilling the task of saving lives (the DQI and Blacksmith insist that fake or substandard drugs and pollution are urgent and deadly problems).[26] This leaves the university lab staff with technologies that are difficult to redeploy in studies that exceed the specific aims of the MQM and Ngagne Diaw projects.

The equipment donated by French partners could soon be reactivated, providing versatile and expandable analytical capacity. Yet as secondhand equipment, its durability and accuracy may be short-lived. And its reactivation is uncertain, as it still depends on the mobility and goodwill of French partners. In 2010, there is still more traffic going the other way: samples and scientists from Dakar going toward French labs to produce data, degrees, and publications. These partnerships do seem to be working well, allowing junior staff from Dakar in particular to obtain and publish results of in-depth studies on topical problems of health risks and consequences of water and air pollution or electronic waste exposure.[27] Yet to do this, they had to wait for scholarships and travel bursaries, creating a sense of suspension and uncertainty, while sharpening the sense of unequal resources and capacity, which might prevent a truly equal working relationship.[28]

THE MEMORY OF DESIGNED THINGS

Enter the lab. Pass through the main campus entrance, over which looms a scrolling digital display from which you read: "The university: a heritage for students, a plinth for the nation. Pfizer, a drug production unit in the service of Africa."[29] On your right is the Faculty of Law, and on your left, the tall lines of the Faculty of Medicine's five-doored entrance, both set behind majestically thick-trunked trees. Keep going, then turn left, until you see a stone bust of Galen donated by Pfizer: you are facing the Faculty of Pharmacy. A short semicircular driveway takes you to the brass-framed glass doors and the stairs up to a central corridor dappled by sunlight filtered through the interlinked patterns of the facade that soften the boxy modernist architecture of the concrete building. On each side of the door are parking spaces identified with chemical and pharmacological sciences: "toxicology," "organic chemistry," "pharmacognosy," "pharmacology," and so forth. These are for the cars of the unit heads, the *chefs de service*, which they often drive over to other ministry of health services—the directorate of pharmacy, the drug control lab, teaching hospital labs, and pharmacies—of which they are also heads.

Turn right along the corridor. An eye-catching logo, designed by a private firm, identifies the door of the main administrative office of the Laboratoire de toxicologie et d'hydrologie (LTH). This used to be a storage space that has been converted into a spacious office, with a corporate-style desk, sofas, and table (thanks, I was told, to one of Dakar's biggest private medical analysis labs). Keep going along the corridor to one of the wings that flank the H-shaped building. The door on your right is identified with a sober plaque of an older professional aesthetic; it leads to other offices of the toxicology unit, and those of the Laboratoire de chimie analytique et de bromatologie (LCAB; recall that these were a single unit before 2008). Straight ahead is the doorway to a teaching lab, set just below an identical one on the floor above; it is spacious and fitted with built-in rows of benches, sinks, gas taps, ventilation fans, and huts.

FIGURE 1.6. Pfizer billboard, main campus entrance, Université Cheikh Anta Diop. Photo by Noémi Tousignant, Dakar, 2010.

FIGURE 1.7. Entrance to the Faculté de pharmacie, Université Cheikh Anta Diop. Photo by Noémi Tousignant, Dakar, 2010.

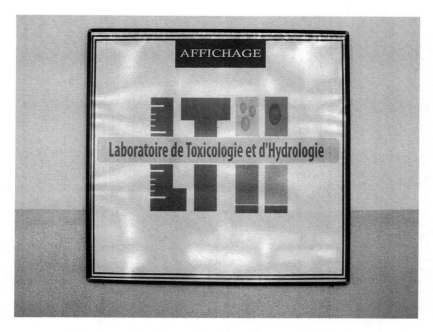

FIGURE 1.8. New logo of the Laboratoire de toxicologie et d'hydrologie.
Photo by Noémi Tousignant, Dakar, 2010.

If the apparatus in the lab bears the traces of foreign investments in its analytical capacity, the aesthetics of its architecture, layout, and signage display other agencies and horizons. The design of the Faculty of Pharmacy bears (and perhaps betrays) former hopes for the production of a modern African professional elite within its walls. Built in the mid-1960s, its architecture is modernist. It echoes the rectangular, mostly low, L- and H-shaped buildings around it, which vary in size and grandeur but not in their basic look. The University of Dakar was created in 1957, on the eve of independence. A wave of public building and urban planning was initiated from the late 1940s by the infrastructural investments of the metropolitan-financed colonial development fund, FIDES (Fonds d'investissement pour le développement économique et social).[30] Unsurprisingly, many buildings on campus, built between the late 1950s and late 1960s, bear a family resemblance with the albeit more imposing government buildings from this period, such as the Assemblée nationale (national assembly) and the Building administrative (known as "le building"), which houses many of the offices of public administration. The modernist angles of the new buildings on the campus of the Faculty of Medicine and IFAN (Institut Fondamental d'Afrique Noire) contrast with the art deco curves of their predecessors, built during the interwar period in the administrative center (a couple of kilometers away).[31]

The faculty building and its built-in labs could be categorized, along with glassware and Bunsen burners, as material remains of French assistance. Yet while apparatus is ephemeral, the buildings and its interior were designed and built to last beyond la coopération. More than their contents, then, their architecture bears past visions of the future, and specifically of a national and sovereign future. Coopérants' teaching and research activities (and their use of lab materials) were temporary by definition: meant to produce the experts who would replace them.[32] The buildings of state administration and education, however, were created as a durable infrastructure of African modernization.[33] The developmentalist policies that made such infrastructure fundable by French taxpayers after World War II (contrasting with the limited and more immediately profitable infrastructure built previously with colonial levies in taxes and forced labor) opened the possibility of African membership in a universal modernity of technical norms and expertise.[34] The creation of an Institute of Higher Education in Dakar in 1950, then of the university in 1957, both firsts for French Africa, concretized this belief in equivalent expertise and convergent development. Students would soon criticize as neocolonial the initial lack of concern with Africanizing university curricula, pedagogical structures, and faculty.[35] Yet as David Mills has described demands for British-equivalent educational standards at the University of Makerere in the 1950s: "[M]imesis was a political principle, not evidence of social inferiority."[36] Both the Universities of Dakar and Makerere became, in the 1960s and 1970s, "objects of national pride," a pride arising simultaneously from their "international" validation of quality and their role in producing qualified civil servants to govern the new nations.[37]

The Faculty of Pharmacy building was inaugurated in 1966, at a time when most of its faculty members and a high proportion (a third to half) of its students were French nationals, and much of its funding came from the French government.[38] Yet soon after, there were demands for a more autonomous pharmaceutical sector and, from Senegalese students and technocrats, for an accelerated Africanization of higher cadres and greater economic self-sufficiency through industrialization. In the report (marked as "for limited diffusion") of a 1969 mission for the WHO, Michel

Attisso, formerly a professor of pharmacy in Dakar, wrote: "[T]he problem of pharmaceutical technical staff is one of the most acute [problems facing the development of the pharmaceutical sector] in Senegal at this time. Technical assistance must only be a step."[39] He counted only fifteen among Senegal's sixty-two pharmacists (excluding academic and research pharmacists), and while the ratio was better in the public sector (six out of eleven), the need was much higher: he estimated twenty-three. To these problems, he asserted, the Faculty of Medicine and Pharmacy offered a "viable solution."[40] He noted a reform of the curriculum was under way to adapt it to "African needs." During this time, studies of the feasibility of state-subsidized local pharmaceutical production were also under way; this project was concretized in 1974. By then, another government program supported the Senegalization of private pharmacy ownership by providing low-interest loans from the state treasury.[41] So while the Faculty of Pharmacy may have been "built by the French," it soon became (and, indeed, was probably designed as) a site for the Senegalization of pharmaceutical expertise by and for an expanding state apparatus.

An "equivalent" pharmacy education took six years, was heavily science based, and concluded with the defense of an original thesis.[42] The ample space in the Faculty of Pharmacy that is given over to the teaching labs displays the importance of acquiring an embodied technical memory of analytical manipulations in the training of pharmacists. In 1995, a pharmacy thesis on practical teaching asserted that the "know-how" generated by lab exercises would be used by graduates "to serve their country by successfully joining existing pharmaceutical structures."[43] By then, such a statement probably seemed idealistic, even outdated: only a handful of the growing number of graduates (the number of students enrolled in pharmacy grew from 742 in 1983 to hover around 1,000 into the mid-1990s) could hope to obtain a government position, or even to work in a lab. Yet at the time when the teaching labs were built, in the late 1960s, the most talented Senegalese and other African graduates were provided scholarships to complete their training in France and were quickly absorbed into jobs of scientific expertise and public authority. The design of the teaching lab, which was once surely state-of-the-art, is a material assertion of synchronicity and equity in scientific education,[44] of the relation between the universality of scientific method (and here, more specifically, of chemical analysis techniques) and the potential access to modernity for all peoples and places. After currency devaluation in 1994 doubled the price of im-

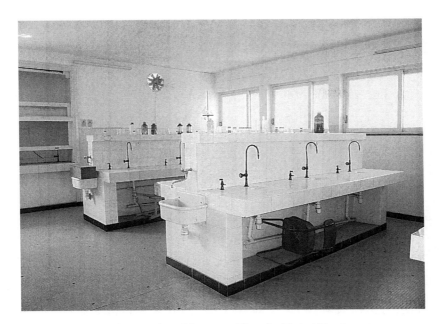

FIGURE 1.9. Toxicology teaching laboratory. Photo by Noémi Tousignant, Dakar, 2010.

ported chemicals, it became more difficult to fulfill the function inscribed in the rooms' design. By 2010, Professor Fall complained: "Practical exercises? We do fewer and fewer. Now, it's mainly demonstrations."[45] Some tests are not even demonstrated, and sessions are sometimes canceled, for lack of solvents and reference substances. One of the lab demonstrators complained to me that she was "forgetting" her skills, that "waiting [for access to equipment and reagents] gives us laziness."[46] Her students lacked opportunities to even acquire the memory of manual operations for detecting traces of drugs, medicines, alcohol, and pesticides in urine and blood (the standard lab exercises since at least the 1970s). The derelict materials within the teaching labs betray their earlier aspiration to the making of embodied memories of universal technique. Instead they index a loss of skill and of memory.[47]

The aesthetic of the university buildings of the late 1950s and 1960s is sober, modern, and solid. It evokes the line from the Pfizer billboard: "the university as plinth of the nation." Yet as the message scrolls on, it reveals that a private multinational firm is promoted alongside nation-building and public service. The billboard seems to announce the hybrid aesthetic created by the lab's new logo, a colorful and entrepreneurial presence amid the worn but durable features of public bureaucracy in the rest of the building. The logo, printed on office plaques and pamphlets announcing the lab's postgraduate degrees, "will make some people jealous," one lab member confides: "Us, we are here, we are working. Others, they don't even show up."[48] What he means by this is that the toxicology lab is, by its enterprising self-promotion and revenue-generating programs, succeeding in a new economy of university education. Those who continue to rely only on old modes of funding will be left behind. Despite the lack of lab activity, Professor Fall keeps busy with teaching, not only in the regular undergraduate program (with its many theses to supervise) but also in the newly created diploma and masters' programs (in risk assessment and quality management, and in biotoxicology). Authorized by recent university-wide reform, these programs are paid for by student fees (rather than state subsidies), a fixed percentage of which the lab can recoup for its operational budget.[49] Fees can pay for consultants to teach on the program, including themselves as well as colleagues from the private sector and foreign collaborators, and for the purchase of office equipment. While the content of the courses is oriented to both private- and public-sector work, it emphasizes a managerial approach to risk, safety, and quality that promises to enhance the competitiveness both of the students who complete it and that of the institutions that might use their skills.[50]

By attending to processes of design, we can find past futures. Here is an aesthetic projection of a common future for academic scientists, the public university, and the state. This future, to be reached via a universally equivalent education and science, was never uncontested, as manifested in the 1968 student protests, which crystallized more diffuse and diverse disagreement over postcolonial pathways to equality and self-determination. There were also more subtle tensions over questions of whom these pathways advanced and benefitted, as the next chapter describes. Still, as designed rather than as remaining, the materiality of the lab is a reminder of the (literal) building of public institutions, with its expectations of autonomy and Senegalization, and of the (re)production of expertise to "serve the nation." At the same time, the very persistence of the lab's old material forms, forms that display an absence of renovation and reequipment, seems to betray the hopes they once held for a place made placeless by universal techniques, and in which the expertise to detect poison and control quality could be embodied.[51] In its current state of obsoleteness and inactivity, the lab is no longer placeless but firmly located on the periphery of global science.

It is also difficult to say exactly how these buildings, amid omnipresent complaints about the decline of the public university as well as the rapidly expanding business of private universities of varying sizes and quality, act as reminders of lost, abandoned, and possible futures. Are they unsettling reminders of greater political commitment to African science, public service, and self-determination (as some older lab members might suggest)? Or (as the Pfizer banner proclaims), do they simply point to the promise of a better future that can be redefined (as "excellence," now a common catchword in university public relations materials) and reached by other means in the current search for entrepreneurial opportunity? Lab members' efforts to animate this space has not transformed it but overlaid it with the aesthetic marks of their entrepreneurship, by which they keep themselves active and attract fee-paying students to the promise of better careers.

I now turn to "keeping" as a third way in which the materiality of the lab is carried toward—projected, remembered, brought into—its futures. With keeping, the focus switches to lab members' actions; to the ways in which they have manipulated, stretched, supplemented, or reached beyond the material capacity they were "given." By keeping things in memory, in action, even in freezers, lab members have also, with foreign collaborators and the state, set and modulated the tempos of scientific work, imagined and moved toward its future, and transformed the past into longer-lasting forms.

THE MEMORY OF KEPT THINGS

Enter the lab. You will soon recognize the shape, size, and fonts of student thesis reports; they are lined or piled up on shelves in every office and stacked—on my first visits, haphazardly, but later in neat chronological order—in the deep drawers of metal filing cabinets. Choose one defended in the 1990s, and perhaps you will find the results of measurements of pesticide, aflatoxin, or heavy-metal traces obtained with the AAS, the GC, or the HPLC at a time when they still worked, but after the end of the projects through which these apparatus arrived. Flip through the pages of the thesis and find a diagram of the analytical system and, further on, a conclusion that calls on "public authorities" to prevent and monitor the contamination of food, water, air, and bodies.

Visit Babacar Niane, a former lab technician, and later a faculty mem-

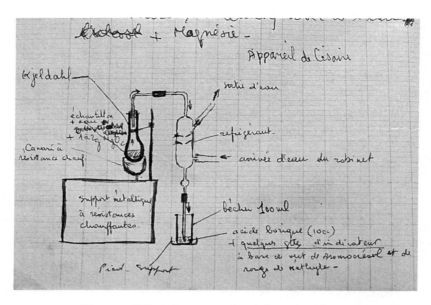

FIGURE 1.10. Diagram of Césaire's apparatus, hand-drawn by Babacar Niane in the late 1960s.

ber, in the pharmacy he opened when he left the lab in the late 1990s. See what he showed me: a hand-drawn diagram of an apparatus invented by his first boss, the analytical chemist Olivier Georges Césaire (whose memorial plaque you may have noticed at the end of the lab's inner corridor); reprints of articles from the late 1960s, crediting Niane's technical help, in which the apparatus also appears; Niane's CV, a list of degrees and internships he proudly points out; and later, in his home, a copy of his voluminous thesis on heavy-metal contamination in fish and seafood.

These documents are part of the lab's archive: the texts produced and kept to carry selected events of its past forward in time. Keeping reports, offprints, and CVs stores and rekindles memories of what could be done with active lab materials. Activating lab materials in the past, as these documents testify, has also meant keeping them *in action*. Lab members and their thesis students kept the machines beyond the end of investments— bilateral, national, friendly—that brought, designed, and left them to remain. They kept them working to generate analytical results that sustained the lab as a busy, functional, and serviceable public institution, one in which pharmacy graduates could become public health scientists, and in which faculty could detect, and express concern about, potential toxic exposures. The lab's archive, then, is of records and mementos, and of the descriptions they contain of keeping analytical materials. We might also add to this archive the samples of urine and breast milk that sit in the suspended time of deep freezers, its "frozen archive," in the words of Warwick Anderson, kept for a future of analytical (re)activation.[52]

The act of keeping is important because it creates traces not of the unpredictable rhythms of activity and idleness set by Northern collaborators (as do remains), nor of expectations of progressive nation-building that have crumbled into entrepreneurially driven fragments (as do designs). Instead, it speaks of struggles for and imaginations of extended tempos of scientific activity and of directed movement that reach for the advancement of knowledge and careers, the regulation of contaminated matter, and the protection of publics. Thus we might see the lab's archive, its records and reports, machines and frozen samples, as made up of—and set in motion by—the ways in which Dakar-based toxicologists and analytical chemists have sought to exercise analytical capacity, as well as of their past and current hopes, worries, ambivalence, or nostalgia.

What lab members kept shaped not only the record to which I had access but also how they guided my reading of it. "Keeping" creates a record of their past action on lab materials but also, especially, one of their past and present aspirations to "good" science. Its narration by lab members in a present of deactivated materiality tends to be nostalgic or hopeful, reaching beyond recent disruptions of activity. Yet it also often evades the past slowness and syncopation of this activity, and the difficult dilemmas encountered in striving to survive and succeed as public scientists. Niane animates his mementos, for example, with accounts of his own moral commitment to forward-moving tempos of methodological inventiveness and professional advancement during la coopération. Of the subsequent time of projects, he emphasizes how machines were used beyond projects to monitor environmental health risks, generating both the possibility of protection and scholarly achievements that opened up and advanced scientific careers. Mounirou Ciss, the Senegalese deputy head of the lab after Gras's departure, similarly intertwines Niane's hard work, his "devotion," with the functionality of machines during this period. These two accounts, which personify the "keeping in action" of functional materials as Niane's virtuous scientific life,[53] underscore the shock of their interruption and loss in the wake of currency devaluation in 1994 and Niane's consequent departure in 1997.[54] Yet they tend to dissimulate the tensions arising from racial/

geographical inequalities in scientific opportunity and already-scarce and precarious investments in the lab from state and foreign sources even before devaluation. While I follow Niane's cues to trace how "kept" rhythms of activity could be tied to individual and collective futures, resisting the fragmenting effects of these tensions, I also seek to discern the difficulties and limited results of keeping, as they are expressed in the material constraints of lab equipment in Dakar and as they are revealed in the modest contents and publication status of scientific reports, as explored over the next two chapters.

The keeping of samples similarly represents an effort to stretch the limits of deficient lab capacity yet ultimately reveals ways in which the absence of flexible, functional, continuously active, and up-to-date analytical capacity in Dakar fragments the time, space, quality, and purpose of scientific work. Lab members freeze samples to suspend them until favors (from private labs, traveling scientists, and heads of overseas labs), competitive scholarships, and biannual travel grants come through to provide transport and access to better-equipped labs elsewhere. Shipping samples is rare and expensive. Many companies, explains Mathilde Cabral, a junior lab member who is doing her PhD work in a French lab that has close ties (and a formal collaboration agreement) with the Dakar unit, refuse to carry dry ice. She can remember only three instances in which samples were shipped, the first paid for by the WHO (a global biomonitoring study of DDT traces in breast milk to assess the impact of the Stockholm Convention on POPs), and the other two by a private medical biology lab in Dakar, to which one of the lab's senior members has links. One of these was for a research project: a pilot study of toxic exposure in a very small sample of car mechanics and battery recyclers. The private lab "helped" with some basic blood analyses and the shipping of blood samples to Lille, which performed analyses of lead levels and exposure indicators. Soil samples were also analyzed for free in another lab in Dunkerque. The second case was for the measurement of lead levels in samples collected in Ngagne Diaw (see introduction), before international agencies became involved in the investigation.

In most cases, however, moving samples entails freezing and waiting. Cabral has carried many of her own samples to the lab in France where she is enrolled as a PhD student thanks to a scholarship from the Agence Universitaire de la Francophonie (AUF, an interuniversity fund for the francophone community). "We keep them at −80 degrees," Cabral explains of the urine and blood samples collected for a study of toxic exposure in people living on and near a large, open-air garbage dump on the outskirts of Dakar, "then put them in small polyester boxes. . . . We simply bring them with freezer packs. When we get there, we put them straight back into the freezer. During a six-hour flight, they don't defrost. We go straight to the lab to put them back at –20 degrees or –80 degrees."[55] She also enlists the help of colleagues. Fall was able to bring a batch of Cabral's urine-filled tubes when he traveled, probably paying for his trip with the biannual study grant (of about 1,500 USD) to which he is entitled from the university as a faculty member. On another occasion, Cabral's French thesis director came to Dakar to teach on the master's course in biotoxicology and carried back some blood samples. Cabral also sometimes transports samples for her colleagues; she brought some water that Amadou Diop, another junior lab member in analytical chemistry, hoped to analyze for his doctoral research in the same lab, and for which he would also apply for an AUF scholarship. Diop used his previous biannual travel grant to carry and analyze a set of water samples collected by a thesis student for the analysis of pesticide residues. Some samples, however, may stay in freezers indefinitely. A student whose thesis research involved collecting the fifty breast milk samples required by the WHO study on POPs doubled the volume of each sample and stored them in the freezer in the hopes of obtaining a scholarship for a master's dissertation in biotoxicology.

For faculty members, Cabral tells me, "ninety percent of research" is located overseas. While, from the late 1980s to the early 2000s, they had been able to keep project machines in action for their (and their thesis students') research, now they must keep and move samples to obtain results. The analytical capacity of the Minilabs can, like the earlier machines, be "stretched" beyond the objectives and resources of the USP drug-quality monitoring project. Some students, usually those hired to carry out sentinel-site sampling and testing, have used the apparatus to conduct their own thesis research projects ("about twenty," according to Diop). These often overlap with the USP target drugs and sites, but some have tested other classes of medicines (e.g., antibiotics or analgesics) or drugs collected outside the sentinel sites. They thus extend, though modestly, drug-quality testing beyond the map drawn up for the sentinel site project. Yet the Minilabs are, in the end, just screening devices. The LeadCare II's capacity is even less extensible: Cabral calls it "semi-quantitative."

And thus Cabral complains: "It's an unfortunate thing to say, but the real scientific work is done over there,"[56] while Diop says of the biannual travel grants: "Every two years, we go there for a month. We take advantage of it to get what we need to make paper," that is, the publications required for academic promotion.[57] The lab's current members inhabit a material space shaped by past, and now terminated, scientific activity, while most of their research work—designing studies; obtaining, preparing, freezing, and shipping or carrying samples; and analyzing these overseas—imprints it only weakly. The uncertainty and postponement of the moment of analysis, as much as its distant location, seem to contribute to the feeling that the "real scientific work" is now elsewhere. Unanalyzed samples signal a stalled research process. The lab's inability to regularly renew glassware and reagents, to replace or reactivate equipment, is associated with the loss of times when research not only moved faster or more regularly but also more fluidly, held together by a set of activities that were predictably tied together over time. These times have been interrupted by moments of waiting; science advances in stops and starts, sometimes slowly and often not at all, and depends on favors, friendships, and the opportunities of projects that travel uncertainly and unequally. By freezing samples, and by cultivating relationships and applying for scholarships, they push back against this fragmentation, connecting Dakar to well-equipped labs in France, to gather moments and places into complete research arcs. In the absence of proximate analytical capacity, keeping in freezers holds together the time of research during the uncertain wait for resources to get scientists and samples overseas, while also creating a dislocation of "real science" from Dakar. Lab members thus manage to create brief insights into the toxic effects of living on a landfill, of breaking open used batteries and electronics, of breathing in leaded gasoline fumes, or drinking water into which agricultural pesticides have leached.

IN RUINS OR AWAITING REPAIR, HERITAGE, OR ARCHIVE? DEFINING CAPACITY IN DEACTIVATION

What is the status of this lab: is it a ruin, in disrepair, a museum, or an archive? What kinds of power—to build and destroy, (re)activate and deactivate—do these forms of obsoleteness entail? What does this imply for how we understand the relationship between capacity and the functionality of materials?

As the place of an abrupt withdrawal of funding and functionality, the lab comes into view as a ruin. Twentieth-century ruins, write geographers

Caitlin DeSilvey and Tim Edensor, are "the structural fallout produced by rapid cycles of industrialization and abandonment, development and depopulation, conflict and reconciliation."[58] In other words, they are products of a power that reverses abruptly, that gives and takes away, builds and destroys (this destruction can be fast or slow), opens and arrests futures. As ruined, the materiality of the Dakar lab—most of it paid for by foreign funding—appears as remnants of fluctuating transnational collaboration that is not palliated by stable national budgets. A common reference on modern ruins is Walter Benjamin, for whom rapid obsolescence is evidence of the illusory promises of capitalism as generative of waste rather than progress and abundance. Has scientific collaboration and capacity-building in Africa been a mirage, with brief and curtailed waves of national and foreign funding producing only incomplete and abandoned infrastructures?[59] Or can we catch in these ruins, as DeSilvey and Edensor see in Benjamin's gaze, a glimpse of past utopian hopes (in the lab's case, for African equal access to a "universal" science)?[60] These questions may ultimately be unanswerable. Yet these two possibilities—that the promise of an innovative, protective, and equivalent Senegalese toxicology (or African science more generally) has been a "trick" to mask or justify the inadequacy of investments in it, or, instead, a hopeful, shared, and motivating ideal—are important to keep in mind.

Ruined spaces call for critical attention to the forces that ruin, that is, to that which produces waste, dereliction, abandonment, and neglect, and defers, withholds, or withdraws the promise of investment. For Ann Laura Stoler and her collaborators, the critical potential of ruins is released by attending to "ruination" as an enduring and violent process of "render[ing] in impaired states."[61] One way in which ruination is exerted is through toxic waste: as mentioned in the introduction, Stoler suggests, for example, that the sludge in Abidjan be seen as a form of "imperial debris." By this she means that Africa is made available and permeable to transnational toxic dumping not just by neoliberal forces of globalization but also by a longer history of exposure and dispossession.[62] Ruination as exposure might be tracked to the site of the lab—a site of potential protection in its material capacity to capture and count such toxic molecules—and how it has been "rendered impaired" by the inadequacy of past attempts to build its capacity. Its current inactive state can be read as a cumulative ruin of these attempts, of the limited and temporary effects of each wave of material investment, and of the vulnerability to ruination produced by unequal terms of participation and access in global science, politics, and economy. In their stillness, the ruins of lab materials might kindle in its visitors an "unsettling awareness of the potency of abject materiality," rendering uncomfortably palpable the experience of precarious and interrupted capacity, and of "irretrievability or of futures lost."[63]

Yet maybe some of the equipment in the lab is not ruined, not yet, but simply out of order. I was told, for example, that the gas chromatograph "still works," but that there has been a problem with its printer interface since 2003 and no funding to bring over the Belgian expert who can fix it. Disrepair may be the result of "slow" ruination by sidelining or "incremental abandonment," but it can also be seen as a default state caused by failure to "fend off" continuous decay by "constant repair and maintenance."[64] Scientific instruments are particularly prone to disrepair, needing constant calibration, adjustments, and often a steady supply of consumables. Simon Schaffer makes a strong case for historical exploration of the labor, relations and "distributions of responsibility" involved in continual efforts to keep instruments in working order. In a short essay, Schaffer focuses on scientific instruments used at a distance from where they were made, revealing the important role of repair know-how and replacement for making science travel.[65] This distance, he suggests, is not a fixed fact but expands or contracts depending on determinations of what it means for scientific instruments to work, on the means available to move parts and tools, and on how the expertise of maintenance and repair is distributed and defined.

A focus on (dis)repair rather than ruination, then, shifts attention away from the sole power of foreign collaborators to "give and take" and toward the ways in which different actors and resources are combined to keep instruments working. For the duration of transnational projects and cooperation agreements, the distance between Dakar and the originating places of apparatus, reagents, and maintenance experts shrank. Equipment did not turn into wreckage immediately after these flows ended. Modest publicly funded budgets, supplemented by the income provided by analytical services that made use of this equipment, could extend the length of their functionality. Yet the distance grew, especially in the wake of currency devaluation in 1994, which halved its foreign exchange value. By the early 2000s, all of the equipment provided between the mid-1980s and the early 1990s had collided with the decline of the lab's purchasing power and been turned into "wreckage."

"Antiques" is what Fall called assorted glassware and built-in teaching benches. Perhaps he meant these are what progressive scientific activity *should* have left behind, and which legitimately belong to the past.[66] Obsolete, they can also be recuperated as heritage: as evidence—to be celebrated, preserved, interrogated, and displayed—of past scientific activity.[67] Heritage, including ruins (especially ruins curated *as* heritage) can offer a glimpse of the past not *only* as foreclosed by ruination, as wrongfully or prematurely deactivated, but also as opening onto former horizons. Professor Fall's "antiques" may carry memories of a time when the lab, with its generically functional modernist architecture, up-to-date lab design, and regularly supplied apparatus, could generate methodological innovation, socioprofessional mobility, and a technocratic elite to serve a young African nation.

Of course, the question remains: why have its worn components not been renovated and the inactive equipment not stored or displayed or replaced? This brings us back to ruination and disrepair: while the building itself is still in good condition, its emptying out of active equipment impedes the functions proclaimed by its design.[68] Still, subtle renovations within the building suggest new functions and futures: office space, furnishings and equipment, and a new logo to promote the opportunities that fee-paying diploma degrees may create in a now highly competitive and commercialized employment market.

Ruins, broken-down equipment, and heritage offer different possibilities for recovering movement. Following Stoler, we can look at ruins not as "dead matter or remnants" but as evidence of *processes* of ruination, both in the past and as an "enduring effect," as well as processes of "strategic and active positioning within the politics of the present."[69] There is also (potential) movement in the (possibility of) repair of equipment, as well as in the former progressive pathways displayed by the destinies and designs of scientific "antiques" read not as useless but as heritage. Stoler's work on colonial archives points the way toward yet another form of movement that might be found in stilled materials: what she calls the "pulse" of the archive. Stoler finds this pulse by looking beyond "archives-as-things," containers of (very partial) evidence about colonial policy and practice, to see "archives-as-process," as made up of acts of recording, storing, classifying, transmitting, reading, annotating, and so on. In these acts, archives become technologies of rule, sites for the exercise of power that are also imprinted by anxieties and expectations about the effects and limits of power.[70] The pulse of the archive is the reverberation of actions by which materials are made into vehicles for transporting bits of the past into the future. Might we find in the lab's materials evidence not only of (de)activation by investment and abandonment, by repair and breakdown, by aesthetic projection and betrayal, but also of keeping things as records and souvenirs, keeping things in place and in action, and keeping things for later activity? Looking to the lab as an archive suggests other possible agencies of activation. It brings into view how lab members have worked within the limits of dependency and interruption to maintain and remember scientific activity, and the value they have placed on its rhythms and directions.

If ruination is a form of power that abandons, sidelines, leaves behind, and dispossesses, archiving, following Derrida, is a "domiciliated" power that authorizes and orders traces within it, that "consigns" in the sense both of "assigning residence" and of "gathering together signs."[71] This contrasts with Edensor's characterization of (some) ruins as sites of material and memorial excess and *disorder*. Interpretation of the ruin's signs is not prescribed and remains unstable; the "pluritemporality" of the ruin is not necessarily managed. The archive, however, in Foucault's oft-cited passage, *orders* "all these things said" in their temporal distinction and coexistence, so that they do not pile up on each other "in an amorphous mass," nor are they sequenced in "unbroken linearity" but "shine, as it were, like stars, some that seem close to us shining brightly from far off, while others that are in fact close to us are already growing pale."[72] By deciphering, reordering, or destabilizing the work of archival power and taking over or displacing its "domiciliation," it may be possible to create, as Elizabeth Povinelli suggests, a postcolonial archive of the "otherwise."[73] By keeping things—keeping machines beyond the end of projects, keeping records and mementos of their use, keeping samples in suspended animation to analyze them elsewhere, later—lab members have also (re)ordered the times and tempos orchestrated by those who built and equipped their workspace. These were and are not radical actions, but they create traces of lab members' working out of their own interpretations and imaginations of the relations between pasts, presents, and futures of scientific (in)activity.

The question of how to capture the movement that once animated and now haunts a stilled laboratory is relevant to questions about *who*—whose power, interests, actions, and resources—*makes and remakes capacity* over time. Is the functionality of materials a straightforward determinant of capacity? Or can we also find a way of approaching their activation that takes into account not only the provision and manipulation of scientific equipment and skills but also the expectations and aspirations in which equipment and buildings are entangled?

To ask how the things of science are paid for, and what these things allow African scientists to do, will by and large elicit stories about transnational flows, via bilateral agreements, intergovernmental agencies, or more recent "global" partnerships (public or public-private), and the ways in which these, often temporarily circumscribed or in erratic rhythms, (re)activate African scientific skills and institutions. Reliance on foreign provisions of scientific equipment may be particularly striking since the 1980s, when most African governments' spending was drastically reduced by austerity measures and economic crisis, and again more recently, as funding for some forms of transnational research (e.g., in global health) has increased. Even earlier, however, since the 1970s, foreign and international institutions contributed high proportions of funding for technoscientific activity and capacity in Africa.[74] This dependence fragments time in different ways. It creates a fractured map of fast and slow-moving spaces of scientific activity. Wenzel Geissler, for example, has described an archipelagic geography of health research in which enclaves or islands of well-funded, world-class science (in transnational collaborative research centers) are surrounded by a "sea" of decaying and underequipped infrastructure.[75] Some government institutions are rendered "static" by lack of funding and equipment, and are unable to participate in increasingly expensive scientific work that requires a spatial and financial concentration of analytical capacity.[76] Fragmentation also seeps into larger African institutions such as universities and national research centers, where some departments or units are successful in attracting foreign partnerships and others are "left behind." "Institutional fragments" that successfully adapt to a new economy of science sit within institutions such as universities that, as described by Roland Waast and colleagues, "once served as beacons of hope . . . [but] have been turned into shells of their former selves."[77] Yet even these successful "fragments" are vulnerable to a segmentation of time by short funding cycles, often interrupting research or making it difficult to formulate cumulative programs.[78] Individual scientists who escape "decrepit" national institutions to gain access to transnational research centers or consultancies with NGOs are also subject to short-term project-based contracts and can at best hope for a succession of ephemeral engagements.[79]

If the breakdown and nonreplacement of equipment within African universities such as Dakar's UCAD has hollowed them into "shells of their former selves," the aesthetic and functional presence of the buildings themselves, the "shells," recall their past as "beacons of hope." The durability of the spatial layouts and architectural designs of colonial and national institutions carry traces of older waves of scientific capacity-building in Africa. The earliest is marked by the political and racial limits placed on the development of autonomous scientific research in Africa: by investing selectively in "technical" services, subordinating the collection of data and specimens to metropolitan analysis, and restricting the training of African medical and scientific assistants. From the 1940s to the early post-independence decades, however, a literal building spree of extended educational and research capacity materialized as university faculties, teaching hospitals, research institutions, and even secondary-school laboratories.[80] These tied the formation of scientific expertise to a future of modernization and, eventually, of nation-building, and have remained in place and in use. Their modernist forms are still legible as expectations of particular pathways (elite and modeled on the metropole, yet making claims for African access and mobility) to global membership and national citizenship. Unlike the equipment that once filled or has been emptied out of these buildings, they still stand, pointing not to precarious flows of resources from outside the continent but to more durable investments, which reached out toward imperial and national futures, toward the metropole, or radiated across national space. They are reminders of longer temporalities and spatial wholes within which scientific equipment once functioned, or at least was meant to.

Capacity is not just given, whether by international agencies, bilateral donors, or national governments, but is also taken, made, and kept by African scientists. This is true of the strategies deployed to attract and maintain transnational partnerships in order to get funding and equipment but also of the ways in which scientists and clinicians deploy available, often-inadequate material and technical resources.[81] Claire Wendland's and Julie Livingston's ethnographies of medical care and diagnosis in a teaching hospital in Malawi and a cancer ward in Botswana are particularly attentive to improvisations and resourcefulness. Such improvisations are not merely creative uses of inadequate or insufficient equipment by which its functionality is stretched. They are also performances of civic, moral, and affective commitment by which, as Wendland writes of recent medical graduates' understanding of having "heart" for patients and for their work, "they *expanded* the definition of what they could give."[82] This situates "capacity" within moral economies and imaginations, defining it not only in terms of technical results (and thus as second best in inadequate material conditions) but also with reference to specific expectations of the duties and obligations of medical care as (public) service and (moral/political) commitment. The functionality of equipment, then, is not simply present or absent. It has degrees and valences, and can be stretched toward goals of empathy, care, information, protection, and expertise. Still, as both Wendland and Livingston insist, and movingly describe, it can only be stretched so far, not only failing to prevent death and suffering (or exposure and ignorance) but also resulting in demoralization and a sense of inadequacy and ethical compromise.[83] Attending only to the debris of foreign investment risks missing local scientists' and clinicians' actions and aspirations. Yet it is perhaps in the frequent futility of their efforts to create "good" science and care with outdated, incomplete, broken-down, and sporadically activated equipment that we can fully discern the violent effects of scarcity and dependence. To stretch capacity is, on the one hand, to reach beyond the technical limits of analytical, diagnostic, or therapeutic apparatus at hand toward a specific result or effect, as well as toward goods including public service and professional advancement. On the other hand, such efforts rarely result in cutting-edge science, optimal therapeutics, or sensitive and comprehensive regulatory knowledge, revealing the limits of agency and of moral imaginaries of care and protection amid lack of material renewal and repair.[84]

Apparatus transferred and left, buildings designed for development and durability, and equipment kept in action are three (among other) ways in which materials have generated capacity for science in Africa. Creating different traces and memories, these actions open onto different possibilities for reading materials as records of action, as imprinted by past and potential movement, and by the forces that set the cadence, duration, frequency, and direction of that movement.

2 · Advancement: Futures of Toxicology during "la Coopération"

"Dakar's location in West Africa," said Gauthier Pille in the inaugural lecture of the chair of pharmaceutical chemistry and toxicology at the University of Dakar in 1963, "imposes a particular orientation of toxicology, especially as some of our students will serve in this luxuriant Africa in which poison abounds and poisoning rites are highly varied."[1] Arrow, fishing, and ordeal poisons had long been targeted by colonial law, anthropological curiosity, and pharmacological discovery.[2] Yet when Pille wondered whether "the poison arrow [had] completely disappeared from Africa," his goal was not, it seems, to evoke social progress, or loss of either tradition or therapeutic treasure, but rather to reflect on the future purpose and advancement of a "tropical toxicology." Tropical poisons challenged classic methods in toxicology and invited novel techniques for analyzing plant matter. Pille urged his students to go forward with him: "I count on *some* of you to develop techniques for isolating and characterizing the active vegetal agents *of your land*."[3]

Likewise, for Babacar Niane, the post-independence decades at the University of Dakar were a time of innovation in analytical methods. Niane was hired at the end of the 1960s, after Pille's death, as a technician in analytical chemistry and toxicology. He began the first of our four conversations in 2010 by describing an apparatus invented by his first boss, the analytical chemist (and brother of Aimé Césaire) Olivier Georges Césaire, which he called "l'appareil de Césaire." First explaining how it worked, he quickly moved on to describe his own qualities as a scientist: the precision of his skill, his commitment to bench work, and especially what he called his "inventive spirit."[4] He recalled how, with Georges Gras, who replaced Césaire in 1970: "*We* tried to develop a very sophisticated technique to analyze lead contents."[5] Niane remembered obtaining the respect of these French technical assistants: he cited Césaire asking, "but how did you do it?" and Gras exclaiming, "you are amazing!"[6] Niane's proclaimed "love" for invention also slipped into a claim for credit: "Every technique," he insisted, "I improved them all!"[7]

Pille's inaugural lecture and Niane's tales of the past both craft accounts of an innovative and convivial postcolonial toxicology. Both evoke close collaborations between French technical assistants and Africans—the students of Pille's address ("some of you") and Niane himself as the inventive technician—in the development of new methods to detect toxins in Senegal. Both suggest ways in which toxicology's innovative qualities attached it to its African location: for Pille, as a response to the social and chemical complexity of African plant-based poisons; for Niane, as the achievement of an African technician. Taking Pille's optimistic and Niane's nostalgic claims to an innovative and Africanizing toxicology as its point of departure, this chapter explores the ambitions for advancement that animated toxicology at the University of Dakar from the early 1960s to the early 1980s.

The sources of information I obtained about the lab during this period are the following: Pille published just a few (he died shortly after his appointment) but clearly programmatic statements on the future of African toxicology, which can be read in light of his colonial career. Niane is my only living witness of this period. His accounts focus on his own skill, creativity, and commitment, and on the recognition he received and deserved, which his narrative locates in the past of subsequent career progress (in the 1980s to early 1990s), then interruption (in the mid-1990s) as an academic scientist. Scientific publications and reports, including theses, as well as some gray literature reporting on visits to the lab and fragments of correspondence, provide some information on its materials and activities during this period. From these can be gleaned, indirectly, a vision promoted by Gras, the French coopérant who directed the lab from 1970 until about 1982, of a Senegalese toxicology that would be equivalent to toxicology in France in its functions and qualities. Gras is also evoked by Niane and Mounirou Ciss, who would replace him as the senior toxicologist in the lab, mainly in terms of his career ambitions. Of the lab's other members, there is little information other than that provided by authorship of theses and articles, as well as the rank and status listed in the faculty's serial publication, the *Bulletins et Mémoires*.

These very incomplete sources contain three—Pille's, Gras's, and Niane's —hopeful, future-oriented narratives, in which toxicology was heading toward improved methods, Africanization, career advancement, and protected publics. Like many accounts of this period, this chapter is in part about the promises of mobility, autonomy, and equivalence for African science that were sustained by public (national and international) funding for science, by nationalist political rhetoric, and by educational policies. In relation to today's recent past, such accounts can be mobilized both as a reminder of a briefly plausible future for African science and to highlight the starkness of its interruption from the 1980s.[8]

Yet this chapter also seeks, by examining tensions within and between Pille's, Gras's, and Niane's stories and trajectories, to highlight limits on advancement through science *even* during this period of relative optimism. If each of their accounts suggests the possibility of shared pathways of advancement—for Senegalese technicians alongside senior expat scientists, of methods of analysis alongside regulatory mechanisms, of former colonial peripheries alongside European centers—they also address, with difficulty and in conflicting ways, the articulation of legacies of colonial science with expectations of professional mobility, technoscientific progress, and sociopolitical modernization. Gabrielle Hecht's metaphor of "conjugation" offers a way of grappling with the juxtaposition of such promises of change (what she calls "rupture-talk") with the continuation of old hierarchies of technoscientific work. Transformative promises, she insists, were not simply negated by continuity. They generated ambitions and expectations that could "shift the tense or change the subject" of these hierarchies even if their (colonial) roots were not erased.[9] I mobilize this insight into the relationship between "roots" and "tense and subject shifting," comparatively, in the stories of three individuals whose positions in an African university laboratory were differently rooted in colonial pasts, and who had different expectations for the future of toxicological work in Dakar. Presenting themselves as both innovators and bearers of an Africanized toxicology,

Pille, Gras, and Niane conjugated the legacies of colonial science with the tense and subjects of announced transitions in distinctive ways, modulating continued coopérant control over toxicology with advancing methods and careers, and with the modernization of poisoning and of its control.

Building on the previous chapter's exploration of the material legacies of la coopération, here I seek to further illuminate the "ambivalence and ambiguities" of ongoing Franco-African relations after independence from colonial rule.[10] As in most other scientific activities at the university and national research institutions in Senegal, expat scientists played a dominant role in the lab (headed by coopérants Pille, then Césaire, then Gras, while many junior faculty members were also French) until the early 1980s. How did the promise of altered futures (technical assistance was meant, by definition, to be transitional) modulate the experience of this present? What, under these conditions, might count as a "Senegalized" toxicology?

As a catchall term, "advancement" both captures and draws together different forms, levels, and subjects of mobility and progression. It can act as a placeholder for the more specific meanings that various actors gave to the transformative promises of science in general, and of toxicology in particular. Advancement also provides a common frame for the different kinds of change—from modest improvements in laboratory technique to modernized patterns of toxic exposure, from rationalized forms of government to steps up a career ladder—that defined the achievements, aspirations, and functions of toxicology in Senegal. The meanings and associations attributed to "advancement" illuminate distinctive understandings of science's Africanization. Africanization might mean promoting the professional mobility of African scientists; orienting scientific practice toward African socioeconomic development; or advancing science through interaction with specifically African environments, expertise, and interests.

Place matters to science, and particularly to scientific innovation, as a located set of imperatives and constraints but also as an imagined location in geographies of circulation and domination.[11] Advancement provides a window onto the difficulties and possibilities of repositioning Africa in global geographies of scientific labor and leadership after colonial independence. Imagining toxicology in Senegal as an innovative science was a potentially radical proposition after French-colonial policies and discourses defined science in Africa as extractive or derivative, and Africans' scientific abilities as merely technical.[12] At the same time, the innovations that were achieved or proposed were modest, and not obviously tied to Senegal's economic and political emergence. Assertions of a regulatory imperative to control "modern" toxins—of the need to monitor, prevent, and protect—also departed from prior imperial geographies of industrial development and protective science. These situated Senegal in a coeval relation—that is, as existing on the same temporal plane—with places that had had a head start in industrial development and regulatory responses to its risks. The tempos ascribed to, and desired for, scientific practice in Senegal indexed complex positions relative to other places of science imagined not only as advanced but also as rapidly *advancing.*

By capturing the meanings of future-oriented toxicology in the post-independence decades, this chapter also brings into view what Julie Livingston has called a "promising moment of critical intersection" between dynamic medical research and a dynamic vision of African health as *already* entering the modern world of complex exposures, including synthetic toxicity and other carcinogenic risks.[13] In retrospect, this appears as a brief moment of possibility for toxicology, bookended by the narrower horizons of colonial and adjustment-era public health and medical science.

CONJUGATING IN THE TIME OF DEVELOPMENT AND COOPERATION

In the 1940s, development programs *began* to soften the scientific hierarchies between European and African, center and periphery, expert and auxiliary, pure and applied that were previously created and maintained by science and education policies in the French colonies. Initially, scientific and technical training was introduced in the colonies to create a subordinate class of auxiliaries to support an understaffed colonial administration. Medical training, including pharmacy and midwifery, was one of the earliest forms of scientific education accessible to colonized populations. Courses were shorter and placed greater emphasis on technical skills than their metropolitan equivalents.[14] Professional mobility via technical expertise contributed to the creation of new indigenous elites but was tightly bounded by the gender and racial distinctions of colonial society.[15] Investment in scientific infrastructure in Africa was constrained by lack of funds (prior to development programs, French colonies were meant to be self-financing) and by metropolitan control over knowledge production.[16] Most "colonial" science was conducted from metropolitan institutions—

for example, in France, the Muséum d'histoire naturelle, the Institut Pasteur "mother house" in Paris, prestigious university faculties, and learned societies—where data and specimens from the colonies were processed into knowledge.[17] While the African territory was conceptualized as an in situ laboratory, it contained few laboratories per se. In imperial circulations of scientific matter, knowledge, and experts, Africa was largely cut out of analysis and theoretical innovation, functioning as a field for the collection of raw data and the application of knowledge.[18] The philosopher Paulin Hountondji invites us to think of scientific and economic colonial activity as parallel extractive processes, as an "analogy between . . . two kinds of subordinating integration."[19]

The French Fonds d'investissement pour le développement économique et social (FIDES), a colonial development fund launched in 1946, mobilized new investments, notably for infrastructure. The ideology of development also sanctioned forms of technocratic expertise deemed "universal" as relevant to the resolution of African social and economic problems.[20] Investments in scientific research and training, at a time of cautious extension of civic rights, loosened the restrictions on socioprofessional mobility for African technical experts. By the 1950s, doctors and pharmacists trained in Dakar were called "African" instead of "auxiliary" and were granted new rights such as private practice.[21] A few also obtained scholarships to obtain full diplomas in France and drop the title of "African" altogether.[22] As independence was announced, then granted, new research institutions and training institutions were created, broadening access for Africans to higher levels of scientific qualification.[23]

In early post-independence Senegal, the figure of the African doctor was, notes Pauline Kusiak, an "embodiment of progress," associating professional and national advancement.[24] Such an association might be subjected to a Fanonian critique of the limits placed by colonialism on imaginations of African knowledge, freedom, and technology. Yet its very possibility was also, as Kusiak points out, threatened in the early post-independence years by the tensions arising between the promise of Africanized scientific work and the slowness of its realization. In formerly French-ruled African countries, technical assistants, known as coopérants, were employed under the terms of cooperation agreements with the French government. Senegal, with nearly 1,500 coopérants throughout the 1960s, had one of the highest concentrations in Africa.[25] Thus hierarchies of scientific work were still ordered by nationality, and race was still a visible marker of status. In 1967, at the University of Dakar, there were still no African professors.[26]

Technical assistance and Africanization were pursued as complementary policies. Coopérants were meant to be replaced by Africans whom they trained or who returned from advanced studies in France. The transitory and transformative goals of technical assistance were rooted in a universalistic conception of technoscientific expertise. This made technical assistants temporary and their political commitment to African development irrelevant. Yet various scholars have identified ways in which this expertise continued to be inflected by place and race, obstructing the "unbinding" of African mobility in hierarchies of technoscientific work.[27] Hecht observed how the advancement of workers in uranium mines in Gabon and Madagascar was tightly controlled as a "socio-technical process" in which "race never dropped out of the equation," deferring the new world order predicated on technological rather than imperial power as heralded by the twin "rupture-talk" of decolonization and nuclearity. She, as well as Guillaume Lachenal in his study of the failed Cameroonization of the Institut Pasteur in Yaoundé and Kusiak in her analysis of discourses on African technoepistemologies, find that moral, psychological, and cultural characteristics were mobilized to define—and racialize as "white"—the expertise needed to move up technoscientific hierarchies.[28]

This racialization of expertise legitimated the slow pace or, in the extreme (but illustrative) case of the Institut Pasteur in Cameroon, the "systematic postponement" of Africanization.[29] Expats benefitted by keeping material and career advantages in both the private sector and in government technical assistance programs.[30] In Senegal, as Rita Cruise O'Brien points out, state authorities also backed the massive and prolonged presence of coopérants, and even opposed putting limits on length of service, by appealing to *quality* and *experience*. Thus what some critics called "technical insistence" was, for Senegalese president Léopold S. Senghor, both insurance against the "cut-rate Africanization" (*l'Africanization à rabais*) that would threaten equivalent standards in Senegalese administration and education *and* a positive valuation of coopérants' durable commitments toward, and accumulated expertise in, an African setting.[31] Indeed, their African attachments could trouble the category of "expat." Lachenal evokes, provocatively, an understanding of "Cameroonization" as applying to expats rumored to have acquired the very proclivities they ascribed to Cameroonians in blocking their sociotechnical ascension. Following Senghor, however, we might also consider how coopérants might have been seen, or have seen themselves, as best placed to make scientific education and research yield advancement for Africans.

SCIENCE AND SENEGALESE MODERNITY

For many of its political and intellectual elite, Africa's self-determination and global membership hinged on mastering science and technology. In other words, decolonization required science, while the decolonization of science was, most urgently, a matter of Africanizing its location, beneficiaries, and practitioners. There existed more radical projects to recognize and develop African "endogenous" knowledge as part of the history and future of a truly global science. Even more radically, some, following Frantz Fanon, rejected the imposition of the concept of the universality of science as a form of colonial violence. In Senegal, Senghor's dominant (but far from uncontested) rhetoric saw "*technicité*" (a technical spirit) as a quality to be imported and assimilated to enable a full, yet distinctive, African participation in modernity. More broadly, Senghor's aesthetic and political philosophy of Négritude, expressed in both poetry and political speeches, explored the nature and possible interactions of distinctively African and European qualities and dispositions. Although shaped by colonial thinking about the essential and thus homeostatic nature of African culture, and widely criticized (Fanon, unsurprisingly, was a prominent critic), Senghor's project was, some argue, deeply syncretic and transformative: "to assimilate without being assimilated." The projected result was a novel aesthetic and personality, which would form the basis of a new civic consciousness and be at once modern and authentically African or Senegalese.[32]

Senghor claimed "discursive reason" and "technicité," even if not endogenous to Africa, as part of a common world heritage to be cultivated by Africans in becoming modern.[33] In 1939, Senghor had written, controversially: "emotion is negro, as reason is Greek."[34] His later writings and speeches nuanced this position, insisting on the capacity of, and imperative for, Africans to "actively assimilate" European and North American scientific progress: by cultivating "discursive reason" to become "men [*sic*] of the twentieth century, in tune with the civilization of quanta and relativity, with the civilization of the atom."[35] This rhetoric echoes other postcolonial leaders' use of science in the "suturing of modernity and nation," notably in the pronouncements of Indian Prime Minister Jawaharlal Nehru or Ghana's first president, Kwame Nkrumah.[36] Like Nehru, Senghor mobilized science as, in the words of David Arnold, a "philosophical and literary pursuit" with authorial power to narrate the "autobiography" of a postcolonial nation.[37]

As in India, however, there were competing claims to an endogenous historical trajectory of scientific modernity. The archaeologist and historian Cheikh Anta Diop was a significant counterweight to Senghor's intellectual influence in Senegalese and pan-Africanist circles. Diop is best known for his tireless quest to prove the "negroid" characteristics of ancient Egyptians, and thus the African origins of modern civilization. He also famously argued that colonization was not the beginning of African history, as the colonizers would have it, but rather its temporary suspension, captured by the expression "colonial parenthesis."[38] Thus Diop's writings about Egyptian achievements in arithmetic and geometry and his urgent call for investment in Africa's scientific future were tied by a historical arc interrupted by the continent's "excessive vulnerability over the past five decades . . . due to technical deficiency." Africa could "once again become [*redevenir*] a center of scientific initiative and decision-making instead of believing it is condemned to remain the appendage, the field of expansion for developed countries."[39] Diop laid out ambitious plans for this transformation by calling for the development of technologies to tap into hydroelectric, geothermal, nuclear, and solar energy sources.[40] For Senghor as for Diop, then, being scientific was not just about being modern as a form of mimicry,[41] but about (re)setting African history in motion: for Africans to (once again) become active transformers of their society and place in the world.

Still, Senghor's government did not invest heavily in specific uses of science and technology to engage social and economic transformation.[42] Achieving a syncretic African modernity was primarily a cultural project for Senghor, as evidenced in the state's patronage of the arts.[43] More significant state investments in scientific research and industrialization were initiated only in the 1970s, after the failure of ambitious programs of rural development (although the latter did feature the diffusion of agricultural technologies including pesticides and fertilizers).[44] Education was also a priority investment area from the 1960s, and programs of mass education sought, by conveying basic technoscientific knowledge and epistemologies, to cultivate a modern civic subjectivity.[45] Senghor's rhetorical appeals to "technicité" were entangled in complex political negotiations between the educated elite, who represented his ideal of rational government, and the rural religious leaders on whose support, through patronage relations, his power depended.[46] As Itty Abraham has pointed out, nationalist discourses about science are more useful for understanding attempts to forge distinctive ideas of the postcolonial nation than they are for situating scientific practice in former colonies.[47]

Drought and student protests in the late 1960s were dovetailed by the weakening of patronage relations and by demands made by a new generation of Senegalese university graduates for positions in power on the basis of technical expertise. As the price of Senegal's main exports—peanuts and phosphates—rose again in the 1970s, the state embarked on an ambitious program of direct public investment in the creation of local industries, creating seventy parastatal firms in the first half of the decade, including units for the production or processing of pharmaceuticals, plastics, and pesticides.[48] Efforts were also made to accelerate the Senegalization of higher echelons of the public as well as the private sector, and to reform university education to make it more responsive to local needs.[49] Amid this general drive to modernize, diversify, and Africanize the Senegalese economy was the creation, in December 1973, of a national coordinating body for research in science and technology, the Direction Générale de la Recherche Scientifique et Technique (DGRST).[50]

The DGRST claimed its origins in intensified government attention to science and technology as key elements of national economic and social development. It tackled problems of oversight and funding, and sought to modify and create research institutions in view of accentuating their "national character."[51] Yet its main domain of intervention was the university, a focus justified by the high proportion of scientists in Senegal, especially of Senegalese scientists (84 percent), who were academics. Lack of research funding and heavy teaching loads made these scientists a "great but underutilized human research potential."[52] Yet Senegalese nationals only accounted for around half of university staff: the other half, occupying most of the higher ranks, were French. Technical assistants still made up 58 percent of all researchers in Senegal (a total of 383) and 12 percent were of other non-Senegalese nationalities. Thus only a third of scientists in Senegal were Senegalese around 1977.[53]

In 1975 and 1976, a special fund to "catalyze" research (Fonds d'Impulsion de la Recherche Scientifique et Technique, FIRST) was allocated by the DGRST. University researchers were the main recipients of its two annual budgets, of fifty and of one hundred million CFA francs. A large portion of the projects funded were public health related, including research on traditional medicines, liver pathologies, leprosy, and cardiovascular disease, while reducing health care costs was also articulated as a priority. Other priority areas were renewable energies, aflatoxins, and social sciences and humanities. Although the creation of a national agricultural research institution (Intitut Sénégalais de Recherches Agricoles, ISRA) in 1974 had launched a "leap in Senegalization," increasing the proportion of nationals in research from 20 to 30 percent in the space of three years, the distribution of FIRST shows a concern to "Senegalize" the content and institutions of research ahead of its practitioners.[54] Thus the DGRST expressed a fairly broad vision of how science and technology could serve Senegalese development, whether practiced by nationals or not, by supporting endeavors across a spectrum of applied and basic research in a variety of disciplines. It did not last: an oil crisis and a drop in peanut and phosphate prices led the Senegalese state to seek a loan from the International Monetary Fund (IMF) in 1979, conditional on implementing austerity measures. It seems that the FIRST budget projected for 1978/1979 was never allocated.

GAUTHIER PILLE: TROPICAL TOXICOLOGY IN TRANSITION

In 1963, Pille called for toxicology at the University of Dakar to turn toward Africa because of the university's "location." This may have been true geographically (although most of the toxic plants Pille mentioned did not grow around the city). Yet socially and institutionally, the University of Dakar from which he spoke was rather ambiguously located "in West Africa." As seen in chapter 1, an overriding concern in teaching and research was to maintain compatibility between this and French universities: students' records show that many pharmacy students transferred to France to complete degrees started in Dakar, while the majority of coopérant faculty members were on limited term contracts and would need to secure positions back in France.[55]

It is from this ambiguously African location that Pille insisted on "tropicalizing" toxicology. Some of his colleagues stuck to teaching and writing about "classic" methods of analysis, supposedly applicable everywhere.[56] Part of the explanation for why Pille instead focused on innovation and tropicalization can be found in the colonial career he describes at length in the biographical preamble of his inaugural lecture.[57] In its text, we read of a passion for tropical toxins emerging over the course of a typically mobile, versatile, and scientifically inclined career as a French-colonial pharmacist (like most, he was a member of the military medical corps).[58] Pille's scientific curiosity was kindled by multiple, overlapping duties in diverse postings across the empire, which took him from Madagascar to Chad, Indochina to Senegal, via Marseilles, where colonial health officers trained and conducted research. He accumulated scientific qualifications and achievements in fields ranging from hospital biochemistry and nutrition research to wartime chemical production. Pille's curiosity extended to African poisons: during his first posting in Madagascar, in the mid-1930s, Pille experimented with fruits of the tanghin plant, used in ordeal poisons (substances drunk by the accused in sorcery trials).[59] By the end of the decade, he was in Chad, subjecting frogs to an aqueous maceration of arrowheads (presumably coated in a hunting poison).[60]

As Pille inaugurated the chair in pharmaceutical chemistry and toxicology (created, he said, at his own suggestion), he described the science of poisons as uniquely suited to pharmacists' scientific versatility. This statement could have been made by a French, metropolitan scientist: in France, toxicology was first taught as a course, then as a specialization, in faculties of pharmacy, while versatility has been a key feature of French pharmacists' practice and identity.[61] Yet Pille's own biography traces another historical relationship between a pharmacist's multiple roles, versatile skills, and toxicological expertise, one mediated by a specific set of interests in toxic properties and contamination under colonial regimes.[62]

Pille's experiments in Madagascar and Chad were probably his own initiative. Yet colonial administrations encouraged, especially from the 1930s, the chemical exploration of indigenous plants, minerals, and products, which were seen as a source of potential import-substitutes or of exportable commodities such as medicines, chemicals, lubricants, fuel, and soap. Survey missions were commissioned, and these often targeted plants with local reputations as toxic. Thus, in the colonies, the association of toxicity with powerful chemical activity was taken as a shortcut for locating potentially lucrative substances in vast, unexplored indigenous flora and technologies.[63] French-colonial science made toxicity a target of extraction, both chemical, in that it led to the isolation of active principles, and economic, in that it could identify raw materials for exploitation and export. Pille and his colleagues might well have understood experiments with ordeal and arrow poisons along these lines. Thus it is no surprise that, in 1963, he spoke of African toxic plants as a cache of valuable chemical activity: "We are all lucky, you especially, that your country still dissimulates therapeutic treasures."[64]

The conception of local poisons as chemical treasures was intertwined with their identification as dangerous indigenous practices that were also the target of colonial law and justice.[65] Pille may have first learned about ordeal poisons in Madagascar because one of his duties there was the provision of medicolegal expertise. He was later brought the remains of a woman killed by an ordeal poison when he was head of the biochemistry lab of Dakar's main teaching hospital in the late 1950s.[66] Detecting toxins would also have been one of his tasks as head of the fraud control laboratory in Madagascar. In these colonial labs, edible commodities were checked for conformity and contamination, and sometimes indigenous therapeutic substances were tested for toxicity. In both fraud control and legal expertise, poisons were not primarily conceptualized as threats to public health but as threats to *public order.*[67]

Pille's publications from the late 1950s focus on this dual dimension of "tropical" toxins as deadly weapons and as indicators of chemical treasure, and thus defined tropical toxicology as judicial and extractive.[68] We can imagine these two qualities also characterized the course on "toxicology applied to the tropics" that he taught at the Institute of Tropical Medicine and Pharmacy at the University of Marseilles in the first half of the 1950s. By this time, in metropolitan France, toxicologists were employed as experts in government committees for regulating chemical safety, as well as in industrial testing and monitoring of toxicity.[69] The absence of industrial control and public monitoring of toxicity in Pille's vision of tropical toxicology can be seen as the result of a combined colonial legacy of weak investment in local industrial development (and thus little perceived need for quality and environmental control), an extractive interest in African toxic plants and products, a judicial concern with indigenous poisonous practices, and health policies oriented toward endemic and epidemic disease. As a course taught in the later years of pharmacy training, toxicology was only introduced in Dakar with the extension of pharmacy education to a full four-year degree.[70]

Pille's conception of toxicology was, to some extent, modulated by decolonization. First, in hinging chemical discovery to methodological innovation, Pille suggested that African plants, and scientists, could alter the ways in which toxicologists everywhere produced knowledge. This departed from a dominant conception of colonial science as metropolitan theory and methods applied to colonial raw resources.[71] And by selectively summoning African students to join his quest, Pille tied methodological innovation to other understandings of Africanization as joint authorship and ownership. Yet the vagueness of his statements on this issue, and his untimely death, left unanswered questions about his views on the sharing of profit and credit.[72]

Secondly, as a hospital biochemist and toxicological expert, Pille also came to see the signs of imminent transitions. In one article, he moved from the "typically tropical expertise" of ordeal poison cases to a sudden upsurge in cases of suicide by chloroquine. In these he saw, "again, a tropical character," both because chloroquine was an antimalarial and because the suicides were linked to white women's trouble adapting to expatriate life.[73] Yet there were also signs of a broader "African" toxicological transition: While colonial law had practically eliminated poison ordeals, Pille noted that synthetic chemicals were spreading rapidly in modern agriculture and public health campaigns to "even the most remote villages."[74] Even Africans, it seemed, had begun committing suicide by ingesting synthetic chemicals.

Until then "unheard of," the issue of "African chemical suicide" connected Pille to other medical researchers at the University of Dakar keen to observe the rapid transformation of Senegalese lifestyles and biologies in the early 1960s. With doctors from the clinical medicine lab, Pille co-authored the first case study of African suicide by aspirin.[75] The same group reported on atherosclerosis, stating: "Like their political status, the pathology of Africans is in transition."[76] These articles announced African pathologies "of the future." Elsewhere, Pille argued against "African norms" of blood chemistry, stating that equivalent socioeconomic conditions would eliminate race-based biological differences.[77] In this, Pille adopted a transition genre common at this time for describing the sociological and biological consequences of urbanization and industrialization, used by psychiatrists, sociologists, and medical researchers who staked Dakar as an observatory of African adaptation to modern life.[78]

This transition model of health would later, Julie Livingston has argued, draw attention away from Africans' exposure to toxic risks. As both prescriptive and promissory for the unfolding of public health, transition thinking has oriented research and interventions in Africa toward "pre-transition," mainly infectious, pathologies, in the goal of moving along a predefined pathway. Thus, Livingston argues, Africans have, in past decades, been addressed as "biologically simple publics," stuck in the initial stages of the transition, for whom cancer, defined as a "disease of civilization," has little relevance. Yet Livingston also identifies an earlier "promising moment of critical intersection" for African public health and toxic exposure in the post-independence decades. At this time, research conducted under the aegis of the International Agency for Research on Cancer (IARC) addressed the specificities, and complexity, of African carcinogenesis as the product of both toxic and infectious exposure, thus operating outside the constraints of a developmental teleology of health.[79] In post-independence Senegal, it was, instead, transitional thinking that proposed to bring toxic exposure in intersection with African public health. Like Pille, his successors at the university, François Fauran and Georges Césaire, adopted a transition genre to describe "intoxications in Senegal" in 1971: "whether he lives in the bush or the big city, the Senegalese man [sic] does not escape social and economic evolution."[80] If, as Livingston shows, a later view of African health as a stalled transition would make toxins invisible, then this earlier view of a transition in motion, as imminent and inevitable, could instead bring toxins into view. Still, tying toxins to modernization placed limits on the intersection of toxicology and public health, for presumably setting up mechanisms of public protection could wait until the magnitude of toxic exposure grew. In none of his articles on tropical toxicology published between 1959 and 1963 did Pille mention the toxicologist's role or responsibility in preventing toxic exposure or mitigating its effects on a population level.

GEORGES GRAS: CIVIC TOXICOLOGY AS COEVAL SCIENCE

Gras arrived in Dakar in 1968, fresh from the defense of his doctoral thesis in pharmacy at the University of Montpellier.[81] He stayed for fifteen years. Unlike Pille, Gras left no programmatic statements about the nature or function of toxicology in Senegal. His vision and ambitions can be gleaned from two sets of sources. The most visible is made up of about twenty texts, coauthored by colleagues in the lab, published in European journals of toxicology, pharmacology, and biology. This rate of international (as opposed to West African) publication was unmatched before or after his tenure in the lab's history.[82] Most of these articles describe improvements in techniques for detecting and dosing traces of toxins, but give little information about the specific uses of these techniques in Senegal (these were some-times described in local—African or West African—publications). Like Pille's project and Niane's memories, they depict the post-independence decades as a time of analytical innovation.

After taking over as head of the LCAT following Césaire's death in a plane crash in 1970, Gras and his deputy, the analytical chemist François Fauran, described spectrophotometric methods for dosing traces of lead in animals exposed to dibutyl lead diacetate in the prestigious *Annales Phar-maceutiques Françaises* and in *Analusis*, an analytical chemistry journal.[83] Yet it is in only *Médecine d'Afrique Noire* (a French-language medical jour-nal established in 1953 for French-colonial Africa) that we learn about the metallic compound's potential economic importance as a mass antipara-sitic treatment in "tropical" chicken and cattle farming, on which Gras had been working for the previous decade.[84] With their junior colleague Cath-erine Pellissier, Gras and Fauran then introduced a colorimetric method for dosing alpha-chloralose in urine that was three times more sensitive than a standard method. European audiences were informed of the meth-od's usefulness for comparing the effects of the molecule's different iso-mers in rat urine, and for detecting criminal poisonings through human urine, but not of how alpha-chloralose might be used in Senegal to poison rats (it is a common commercial rat poison) or to control bird pests (a ma-jor problem for Sahelian agriculture).[85] Fauran had previously developed a simple method for dosing postmortem metabolites of chloroquine, but his readers would not learn about the problem of chloroquine suicides.[86] With another junior lab member, Janine Mondain, Gras published a series of articles on the dosage of mercury and methylmercury. In the more pres-tigious pharmacology and biology journals, they presented a method for studying mercury bioaccumulation in fish and a novel colorimetric tech-nique developed specifically for use in even "the most modestly equipped" of "developing-country laboratories."[87] In specialized toxicology and med-ical oceanography journals, they reported the results of mercury analyses in Senegalese fish, blood, water, and hair but did not emphasize the eco-nomic, environmental, or public health significance of these tests.[88]

This publication record locates Dakar as a place of possible professional and methodological advancement but does little to place these techniques within emerging economic and regulatory practices in Senegal. Gras's methodological work obtained promotions and professional opportunities: tenure in 1973, a professorship in 1976, and an academic post back in Montpellier when he left Dakar in the early 1980s. For Gras and Fauran, staying in Dakar was not an option. Senegalization was slow but inevitable. By the mid-1970s, they knew that two bright Senegalese students had begun their doctoral studies in toxicology and analytical chemistry in France. Creating a smooth career arc between Dakar and Montpellier depended on maintaining scientific synchronicity between them. Modest analytical innovations allowed them to connect and keep pace with, and eventually return to, French toxicology and analytical chemistry.

In the binders Gras left on the shelves in his office when he returned (some say precipitously) to Montpellier, still there in 2010, I found a second set of records of his fifteen years in Dakar. Fragments of correspondence, unpublished reports, and some catalogs and bills for scientific equipment, combined with some publications in African medical journals, offer a glimpse of Gras's efforts to position the lab as part of emerging African regulatory routines and economic development projects.

To the director of the research commission, from which he hoped to get funding for a mercury analysis project with Janine Mondain, a junior lab member and doctoral student, he invoked as a precedent "the research we have been doing for the past four years on environmental pollution by pesticides."[89] This pesticide research could also have been framed as protecting economic interests: the FAO had funded two thesis projects, conducted by two junior lab members, Pellissier and Boubacar Cisse, as part of a project on the control of avian pests that was elsewhere linked to the economic imperative of agricultural intensification.[90] So could the mercury project: to the Maritime Fisheries Agency—from which he hoped to obtain free fish—Gras pointed out that mercury detection was "directly linked to economic issues of interest to Senegal."[91] At the same time, Gras was able to persuade consultants for a UNESCO mission on an African network of environmental chemistry in 1976 that his "institution seemed to be really interested in pollution problems."[92] He outlined plans to detect traces of pesticides that were banned but still in use, thus expressing his concern with problems posed by contraband products and lack of access to information. Gras also volunteered the lab to be visited by an FAO mission in 1979, which was to assess Senegal as a potential host for a regional center for the control of food contamination; he suggested it had the analytical capacity to lead a program to monitor fish contamination.[93] Thus Gras argued that toxicology was needed in Senegal to monitor intensified production in fishing and agriculture, and, more generally, the potential contamination of environments and foodstuffs by these and industrial activities.

Poison control was another area in which Gras asserted toxicology's relevance to public protection. Following a collective poisoning by hydrogen sulfide in 1973, Gras reportedly pressed public authorities to create a poison control center. Approved in principle, the center's creation was still announced as imminent in 1984.[94] Poison control centers, as Robert Broadhead has argued, refashion poisoning from individual danger into a *social* problem, requiring expert management on a collective scale. Although different logics and timings might subtend the establishment of poison control, from the problematization of accidental childhood poisoning in the 1950s United States to collective poisonings in Senegal in the 1970s (and, again, as we will see later, in the 2000s), they draw toxic risk into the arena of public health. In France, where specialized centers were established in the 1960s, these not only enabled emergency responses to intoxication but also the collection of epidemiological data and the formulation of preventive strategies.[95] For Gras, a dedicated center could institutionalize public services the lab was already performing by responding to numerous requests for analyses from state hospitals facing toxicological emergencies and from the police investigating suspected poisonings.

Gras's record is also striking for its silences; it contains no announced program for Senegalese toxicology or any discussion of the distinct challenges to a science of poisons in Africa. Neither modernization nor Africanization is explicitly broached. More puzzling is the lack of explicit association between innovation (the analytical methods he and his colleagues worked on) and regulation. Yet they apparently sought to develop "low-tech" techniques that could serve robust and inexpensive analytical and regulatory capacity, able to withstand the difficulties of maintaining instruments and laboratory supplies far from where they were made, and with modest budgets.[96]

Still, in his separate assertions of the lab's capacity to innovate and to regulate, as well as in his omissions of the particularities of its African location (except in the preference for low-tech methods), Gras implicitly presented, or at least allowed for the imagination of, Senegalese toxicology as existing on the same temporal plane as toxicology in already-industrialized locations. This implied Senegal's existence in modern times, as capable and deserving of an innovative and regulatory toxicology, rather than as lagging behind in the transition toward modern forms of exposure and government. In other words, that Gras found it natural to plan for greater toxicological control, asserting the vulnerability of Senegalese foods, environments, and bodies to exposure (whether or not they were already exposed), can be seen as the assertion of a coeval relation with the already-industrialized, contaminated, and (at least partially) regulated world (even though some of these regulatory mechanisms were quite recent).

Coevalness made it possible for Senegalese toxicology to be a "civic science," in the sense described by Kim and Mike Fortun. Gras proposed a toxicology that was both innovative and regulatory during a time, in the 1970s, of expansion and self-conscious modernization in the Senegalese state, which also followed on the heels of tightened and extended regulation of industrial and toxic hazards in Europe and North America beginning in the 1950s. By expressing interest in "pollution problems" and proposing the creation of a poison control center, Gras suggested—without making a fuss about it—that recent regulatory concerns and institutions could, with little lag, take root in an imminently modern nation.

This proposed coevalness was a departure from colonial extractive and judicial concerns with toxins, and with his predecessors' emphasis on transition as the precondition for modern and generic, rather than tropical, toxicology. Yet it was brief. The time of state expansion, and of reliably if modestly funded university science during la coopération, was curtailed by cuts in public budgets for both innovative and routine analytical activities. This combined with technological changes—already under way in the postwar decades with the emergence of automated analytical systems and genetic toxicology—which made cutting-edge science more expensive and thus less accessible to modestly funded public institutions.

Furthermore, the radical potential of Gras's proposition of coevalness seems to have been dimmed by his failure to address the colonial past and aspirations to development as anchors to claims to African advancement.[97] In envisioning a "universally" civic and modern toxicology for Senegal, Gras apparently avoided the issue of the Africanization of its practitioners or the modernization of its publics. Perhaps this made it difficult for Gras's Senegalese successors to appropriate his regulatory projects, although reduced expectations of state action from the 1980s also certainly played a big role. Neither Mounirou Ciss (who took over the top toxicology position from Gras) nor Doudou Ba (the analytical chemist who succeeded him as head of the lab) remember much about Gras's projects for environmental monitoring and poison control.

Though Ba vaguely mentioned to me that poison control had been suggested decades ago, he dismissed the research conducted before his arrival as insignificant, suggesting it had not really been relevant to Senegalese priorities. What Ciss remembered about Gras, whom he replaced in the early 1980s, was his abrupt departure and a sense that he abandoned Dakar. Ciss told me in 2010 how, just as he returned from France, Gras was offered a job in Montpellier.[98] "He absolutely had to go and there was no way of keeping him. That's how I understood it. He left overnight. So I found myself on my own, with no guidance. . . . I replaced him automatically. . . . I know the authorities weren't happy because of the way in which [he left] . . . and he himself was mad at everyone because he wasn't decorated, although in reality it's the way in which he left. Because his future was there, there was a position so he had to go." But then Ciss shrugged and said: "He had no choice." Did he mean that as an expatriate, Gras would have to leave anyhow? Or that he could hardly be blamed for taking up better conditions for scientific work?

For Ba and Ciss, perhaps the citizenship of scientists, and their specifically African/national ambitions, were key requirements for a truly civic science.

BABACAR NIANE: AN INVENTIVE AND AFRICAN TECHNICIAN

Niane first described for me his past as a lab technician in the office of the private pharmacy he opened after leaving the university in 1997, which had generated enough profit for him to build and furnish a three-story house in the "middle-class" neighborhood of Liberté 6, and to send his children to university, including one to study pharmacy and another medicine. By the mid-1990s, with the lab's supply budgets and salaries reeling from the 1994 50 percent devaluation of the currency of the Francophone African Community (CFA), Niane felt his university position could no longer sustain a progressive career trajectory and family life. By then he'd managed to rise to the level of *assistant*, an academic post with lecturing responsibilities, but still a junior one.[99]

The pride and pleasure of innovating that Niane remembered had, by the time we spoke in 2010, been disconnected from his career and livelihood. They were closed off in a past of greater possibility for both professional and methodological advancement. He described la coopération as a time of regularly renewed glassware and Bunsen burners that he could, as an ambitious and talented technician, manipulate with dexterity, vigilance, and creativity. Of course, Niane was nostalgic for a time of greater means and mobility in science, and his nostalgia for innovation told me as much about his later experience as it did about the lab in the 1960s and 1970s. Yet in the expression of this nostalgia, I think he also captures a central tension in African experiences of scientific work in those decades, when a new sense of possibility opened up by technical skill and talent rubbed against the resentment of occupying the lower levels of a hierarchy still largely and visibly ordered by race.

Inventing, in Niane's account, is the tactile and aesthetic pleasure of manipulating elegantly shaped glassware, which he sketches out for me to explain how Césaire's apparatus worked. It is also a talent, and he takes pride in its recognition: "Personally, I improved many techniques. I'm very curious. . . . Césaire would ask me: 'But how did you do it?'" It is a disposition: Niane proclaims to me his love "of invention" and "of discovery," his curiosity and "inventive spirit" as proof of scientific virtue. Allied with vigilance, precision, and hard work, this love and spirit testify to his commitment to constant, incremental improvements in analysis. "The real pharmacist," he tells me, "is defined by the bench, by work." He opposes benchwork not only to commercial work but also to his later (from the 1980s) work with analytical systems (e.g., the gas chromatograph); these machines, he says, "make you lazy." In contrast, manual methods were malleable, always prone to error, but also improvable, inviting a constant search for greater ease and precision.

The core strand of Niane's narrative intertwines this methodological quest with a personal one. In our third meeting, he follows his usual declarations—"I love invention . . . running the lab properly"—with the description of his recruitment to the university. Césaire recruited Niane on the basis of publications he had contributed toward as a technician at the ORANA, a nutrition research organization. Niane accepted the position because "the university environment was attractive. I wanted to develop myself. . . . I had ambition." Niane's qualification as a technician was a vocational high school degree obtained at the Lycée Delafosse, an institution founded in the late 1950s, proclaiming itself a source of future-opening technical skills for Africans.[100] After joining the lab, Niane started studying for an academic high school degree, the *baccalauréat*, by taking night classes at Lycée Saint Michel, a private Catholic secondary school. He then tried and in 1980 succeeded in passing the entrance exam for the Faculty of Pharmacy: one of only two Senegalese among eighteen admitted that year.[101] He registered in the pharmacy course and took nearly a decade to complete his degree by working nights and often sleeping in his office at the lab. His graduation thesis is a hefty tome that far exceeds the requirements at this level; he gave me a copy and said he still listens to an audio recording of its defense. This is no minor achievement for a man who began his primary schooling at the age of eight.

Césaire and Gras, Niane told me, had encouraged him to pursue his studies. These citations of support echo Niane's quoting of their "amazement" and other expressions of appreciation of his skills. Yet while little bitterness clouds the dominant note of triumph and determination in Niane's narrative, there are hints of what seems obvious by looking at the publications and career trajectories of his superiors. His contributions were recognized as "technical," as skillful rather than creative or intellectual, and credited as such in footnotes or in smaller print under the main authors' names. Although Niane was on his way "up," the techniques he claims to have co-invented had little effect on his own career progression, contrasting with their very clear impact on Gras's rising status. Only once did he allude to this disjuncture, when he told me: "With Gras there was no slack. Because he had to apply for tenure. I'm the one who did everything, especially with lead and mercury."

Was Niane, then, an "invisible technician" as described by Steven Shapin: a scientific laborer whose work might be recognized as skillful and even essential but who is denied the privileges of authorship and visibility?[102] Shapin explains that technicians are made invisible in the political economy of science by broader logics of social and political exclusion. In early modern England, technicians were servants and were excluded from the franchise. Others have explored how, in other historical contexts, race and gender identities have limited the crediting of creative, nontechnical contributions that might make scientific workers more visible.

In early post-independence Senegal, laboratory assistants had recently been colonial auxiliaries—a status that did not lead to the higher ranks occupied by the French. Even Africans who were trained in medicine and pharmacy were qualified as auxiliary or, from the 1950s, as "African," conflating racial identity with "merely" technical expertise. Only in 1962 was pharmacy education in Dakar raised to the level of a French state degree, and yet this expertise—with a high number of students being French—was still unevenly accessible along racial lines. Only a handful of Senegalese students obtained bursaries for health or scientific studies in France beginning in the 1950s, or admission to the new Faculty of Medicine and Pharmacy in Dakar in the 1960s. Despite his scientific talent, ambition, and curiosity, it was unlikely that, by the late 1960s, Niane could have become anything other than a technician.

Yet if we consider how Niane saw, and now remembers, where his talents might lead, his position was quite different from that of a colonial auxiliary. At the same time that Niane worked with Gras, Ba and Ciss were completing PhDs in analytical chemistry and toxicology in Paris, having chosen their specialties to fill the vacancies anticipated with the departure of the coopérants.[103] The fluidity of professional mobility during a time of "category-jumping" for educated Africans, as Frederick Cooper has said of the post-independence decades,[104] meant that Niane might imagine that he would someday, through the virtues of hard work and commitment to scientific advancement, attain a position parallel to that of these French men who officially claimed full authorship of the methods he saw himself as having helped to invent.

Thus Niane could reconcile his subordinate technical status and lack of visibility in print with a remembered contribution that was *both* dexterous and inventive. The qualities associated with repetitive gestures of precision versus the ability to create novelty can generate tension, particularly when they lead to different levels of credit and status inside and outside the lab. In his article on the brilliant black technician Vivien Thomas, Stephan Timmermans shows how such tensions could produce "a torque outside the laboratory" in the segregation-era United States. Thomas was recognized as both highly dexterous and intelligent inside the lab but lacked status outside it.[105] In other cases, the public representation of technicians highlights skill at the expense of a creative, intellectual contribution. For Niane in post-independence Senegal, however, expectations of advancement seem to have mitigated the potential tensions between, on the one hand, the credit he felt he deserved for *both* his technical skill and inventive abilities and, on the other hand, the limited visibility of his contributions as purely technical. Timmermans asks of Thomas's case: "When others routinely take credit for one's work, can technological pleasure compensate for racial and class subordination?"[106] I suggest that in Niane's case, the pleasure of skillful and creative technical work was amplified by potential mobility and its associated work ethic, compensating to some extent for the "merely" technical justification for his subordinate status and ambiguous recognition.

Applying Hecht's notion of conjugation to Pille's, Gras's, and Niane's hopes for advancement can untangle some of the promises and ambiguities of toxicology at the University of Dakar in the post-independence decades. It also comes back to Itty Abraham's observations on how differently scientists can experience a "same" postcolonial place: "Postcolonial locations thus include relations of weakness and possibility, valences that cannot be known in advance but that are products of historically situated intersections of the political economy of place and unequal location within transnational circuits of knowledge flow." Shifting from discourses on the trajectory of nation and science to located "everyday" scientific practices, Abraham argues, obtains understandings of place as "uneven and unsettled . . . where location no longer offers a one-dimensional and stable reference to knowledge."[107]

Following Hecht's metaphor: what were the roots, subjects, and tenses of the various conjugations of toxicology's future in the lab? Pille conjugated an extractive toxicology in a new person by inserting the subject "we" that included "some of you," that is, the African students who would accompany him into an unexplored, toxic, and thus rich terrain of chemical discovery. Together, they would take up the challenge of African toxins by pursuing analytical innovation, thus envisaging a transformed tropical toxicology. Pille's practice of toxicological expertise, as well as biochemical analysis, also changed the tense of toxic exposure; no longer a remnant of African poisonous traditions, it was announced as an African pathology of the future. This was a toxicology of the tropics and of transitions, bound to the distinctiveness of African natural environments and cultural traditions that were on their way toward, but not yet arrived at, a time of modern risk and protection.

Gras does not seem to have considered colonial science as the root of post-independence toxicology. This opened up a radical possibility: that of an African toxicology that advanced synchronously with toxicology elsewhere (e.g., in the former metropole of France) by improving techniques and initiating regulatory mechanisms, regardless of differences in level and pace of industrialization, contamination, technoscience, and government. Yet he did not address the colonial African past as a point of departure for post-independence aspirations to Africanization and development. This ultimately made Gras's own hopes and achievements a forgotten past for the history of toxicology as a Senegalese and civic science.

Niane conjugated innovation by changing the subject of authorship to "we," even adopting the first-person singular to proclaim his inventiveness. The tense of crediting, however, was still in the future. These tenses and persons are now complicated by what has happened since: Niane's subsequent hard-earned degrees and promotions, and the sudden interruption of an upwardly mobile career as a scientist in the mid-1990s. At a time when Niane remembered this mobility with pride and nostalgia, he could give an account of working in a position constrained by colonial limits on education and mobility as the prelude to a forward-moving trajectory generated by talent and hard work. For Niane, being inventive was integral to a sense of scientific virtue, combined with long hours at the bench, constant vigilance, and commitment to precision and improvement, that he still defined himself by but that no longer defined his future.

The ambitions for advancement described in this chapter were articulated by the Senegalese men who stayed (and whom I could speak to) and by the French men who left strong traces in print. They only mention in passing the French women who served as junior staff members in the department, such as Catherine Pellissier and Janine Mondain. Their inclusion as authors (though rarely as first authors) and as academic staff members in faculty bulletins means they left stronger print marks than did the technician Niane. With these, we can follow them making slow progress up the academic ranks but also see how quickly coopérants like Gras, as well as the two Senegalese scientists who replaced him, obtained tenure.

Ciss and Ba defended their theses in September 1978 and were both tenured within five years, even though their only publications were derived from their doctoral work. In contrast, Catherine Pellissier was hired as an assistant in 1967, but despite defending an advanced chemistry degree (the equivalent of a master's) in 1978, she had not, by 1982, risen above the rank of *maître-assistant* (just above the most junior position of assistant) and did not complete her doctoral degree until 1988.[108] Janine Mondain had a similar trajectory. Ba told me this about another female junior lab member, Monique Hasselmann, in order to demonstrate the long-standing ties to Dakar of her husband, a coopérant in analytical chemistry: "[Claude Hasselmann], he's Senegalese. He left his wife and children [when he went to Strasbourg]. . . . Mrs. Hasselmann stayed ten years or more; I called her to the lab when he left."[109] If, for these women, Dakar offered possibilities for (slow) scientific professional mobility (and perhaps of cheaper domestic comfort and "help"), their gendered positions were modulated by the conditions of both expatriation and Africanization that favored their male colleagues. Because they did not stay in Senegal or in Senegalese memory, it is difficult to know what they hoped to get out of their time in the lab. In retrospect, then, we can only read advancement as conjugated in the masculine.

3 · Routine Rhythms and the Regulatory Imagination

Doudou Ba and Mounirou Ciss returned from France, the first with a PhD in analytical chemistry, the second in toxicology, shortly before 1980. Not long afterward (probably in 1982), Ba applied for a promotion. Summarizing his career, he explained that his lack of research and publication activity since returning to Dakar was due to "material difficulties ... (apparatus and reagents), added to administrative and teaching tasks."[1] He and Ciss were soon promoted to full professorships anyhow, probably on the basis of the few but good publications that had come out of their doctoral research, and perhaps also their contributions to teaching and government. There were still relatively few Senegalese with PhDs; many academics, at least in the medical sciences, were cross-appointed to positions in teaching hospitals or as government experts (in laboratories, health policy, etc.). This opened other pathways to successful careers as public scientists. Meanwhile, teaching loads were growing under the pressure of rising student numbers and freezes in public hiring. By the early 1980s, academic staff had little time or equipment for research. Just as toxicology was finally being Senegalized, economic crisis and cuts in public spending were threatening its future as an innovative and regulatory science.

And yet, surprisingly, the two decades that followed Georges Gras's departure are remembered by lab members as being filled with useful analytical work. The main sources of this activity were three substantial pieces of equipment identified as "project machines," that is, as obtained by the lab through project-based transnational collaborations. These machines—an atomic absorption spectrophotometer (AAS), a gas chromatograph (GC), and a high-performance liquid chromatography (HPLC) system—provided the capacity to test for the presence of some of the toxins most likely to contaminate Senegalese foods, bodies, and environments: heavy metals, pesticides, and aflatoxins. Refracted through a present that lab members experience as an absence of analytical capacity, this period of project machines is now associated with the ability to work on issues they identify as being relevant to public authorities and public health. When I asked Ciss to identify the main problems toxicology addressed in Senegal, he first listed accidental and occupational poisonings, especially with pesticides, and then added: "we got interested in heavy metals . . . because in the lab we did have equipment that allowed us to do those kinds of analyses, with the AAS and all that. . . . And there were especially people who were devoted, very devoted. . . . We could really do a lot of things, [and] we supervised a lot of students and all."[2]

Niane also spoke of being busy and useful during this period. No longer referring to methodological invention and improvement, as for the earlier period, he instead recalled contributing to "activities in areas of interest to government, to public authorities, like the pollution of marine organisms, pesticides, the problem of fraud . . . not a month went by when we didn't target a sector."[3] In the 2000s, after Niane left the lab, public salaries (which had lost purchasing power from currency devaluation in 1994) were increased, but, around the same time, the machines broke down, and little money was reinvested in lab work. Niane, who completed an advanced degree in the early 1990s for which he analyzed pesticide residues in market produce, remarked wistfully in 2010: "Before, means were invested in equipment. It was possible to go get a plant and to determine what was in it. That . . . is for the country, for the population. Salaries . . . don't help the population. Before [lab members] complained about salaries, but they had a love of research."[4]

In this chapter, I explore how lab members entangle(d) past analytical capacity with the possibility of keeping busy and of being good, of filling time with hard work and maintaining the lab as a functional public institution during a time of austerity measures and stagnating or deteriorating state institutions. This entanglement was, and is, associated, in particular, with the provision and uses of project machines. This appellation might suggest that such equipment was restricted to the limited time and objectives, defined outside the lab, of international projects. Yet both international investments in the machines, and the ways in which they were used by lab members, made this capacity open-ended in time and function, that is, potentially continuous and versatile, and capable of serving projects but also careers, governments, and publics. Here, I examine how projects (focusing on the marine pollution project that provided the AAS, given the lack of information about the "Italian project" and the fact that Project Locustox mainly invested in capacity outside the lab, as described in chapter 4) and especially lab members tried to make this equipment generative of scientific and regulatory continuity. Associated with the regular testing rhythms of monitoring, the activity of the machines sustained, to some extent, the continued *survival* of a public institution, of public protection, and of public scientific careers against the broken or finite temporalities of economic crisis, transnational projects, and contractual analytical testing. Contrasting with a present of inactivity in the lab, also a present in which lab members' activities have moved elsewhere, these remembered rhythms also stand out against alternative pasts: the possibility that scientific work could have stalled completely in the 1980s, or have been wholly dictated by the "punctuated" rhythms of projects and (private) contracts.

At the same time, however, the memory and imagination of continuous and useful work masks the brevity and scale of the tests performed with the machines, which are conducted on small batches of samples. Despite calls for "regular monitoring" (of food quality and safety, of environmental contamination), this work was unlikely to be picked up by regulatory mechanisms. As both aspiration and memory, the continuous activation of a protective toxicological science by a few pieces of functional equipment was and is a *fiction*. The modest studies performed in the lab mimicked regulatory testing but did not enact or lead to real control of toxic risk. Still, this is a productive fiction, through which lab members keep open the possibility of regulation and of a professional ethos of public service against the decline of public science and higher education, and the possibility of escape toward projects and the private sector.

In Senegal, as in other newly independent African countries, the state, in the 1960s and 1970s, was a maker of futures. Through ambitious development planning, it designed a future political economy of shared prosperity and welfare. The proliferation of institutions of planned development, from marketing boards to health care facilities and parastatal industries, drew an increasing number of people into the state's orbit.[5] Whether animated by ideals of public service or by patronage relations of shared gain, the state's resources and actions intertwined private and national destinies.[6] From the late 1970s, however, the combined effects of drought, falling commodity prices (in Senegal's case, for peanuts and phosphates), and debt (which some attributed to state inefficiency) led Senegal, followed by other African countries, to obtain loans from the International Monetary Fund (IMF) and World Bank.[7] These loans were conditional on the implementation of measures aiming to increase the competitiveness of African economies in international markets. Though "structural adjustment packages" (SAPs) varied, their core measures sought to liberalize markets and reduce public spending, especially on "nonproductive" social services. Not only did states have less money to invest in development, but planning itself as a mode of future-oriented government was under ideological attack, for economies should respond flexibly to market conditions rather than political economic designs.[8]

Many have observed that the effects of liberalization blended with those of fading economic and political optimism, and with the impact of a post–Cold War drop in foreign aid, to usher in a time of "permanent crisis" and decline across Africa. Charles Piot, writing about Togo, has most clearly described the experience of change in the state's presence and resources as a temporal one. The continuous, progressive temporality of the developmental state was displaced, after the Cold War, by the episodic, performative presence of a less resourced state. In addition to this new rhythm of "the political" were other forms of "nonlinear punctuated temporality" brought by the other bearers of hope, services, and development who filled in spaces of state absence, notably evangelical churches and NGOs.[9] Other anthropologists have also described subjectivities of "the crisis" in temporal terms: as experiences of reversal, stasis, and decline; as the loss of former expectations and as nostalgia for the futures they promised; and as a growing inability to plan beyond the immediate moment due to rising prices, unemployment, or delayed salary payments.[10] These observations in part describe the effects of general economic crisis, but these also include the specific effects of SAPS on socioeconomic inequality, purchasing power, and employment opportunities.[11] The erosion of state capacity exacerbated experiences of uncertainty and decline, and certainly changed many Africans' expectations of the state as an employer and a career path; as a provider of services, care, and infrastructure; and as a credible source of historical narratives leading toward better futures. Mamadou Diouf has described the urban youth movement *"set/setal"* ("clean/cleanup") that arose in late 1980s Dakar as mobilizing, against the disinvestment of the state from public spaces, a new sense of historicity to displace the nationalist master narratives of the political class they no longer trusted with their destinies.[12]

Crisis and adjustment did not affect all Africans in the same way. Sectors in which the developmental state had promised the most, and cut most drastically, such as health and education, were particularly affected.[13] The livelihoods of the vast majority of Africans, especially the poorest, became more precarious.[14] Those employed by the state, including most health workers and scientists, were certainly not the worst affected in material terms. Yet they stood to lose not only material security but also the frame provided by the state for defining themselves as expert, successful, and good, in terms of public service.[15] Not all countries experienced crisis in the same way, or at the same rhythm. Although Senegal was the first African country to implement austerity measures, its government managed to postpone some of the disruptions of adjustment through a combination of "disobedience," ability to attract donor funding, and membership in the franc currency zone.[16] By the early 1990s, however, it started intensifying reductions in spending through cuts in public salaries and privatization. In 1994, the CFA currency was devalued (after other African currencies had already experienced rapid inflation and devaluation), drawing those who depended on it into what Achille Mbembe has called "a new time of the world."[17] This sudden 50 percent drop in purchasing power was felt acutely in both public and household budgets, doubling the price of imported goods overnight.

Already in the 1980s, as Diouf observed in his study of "set/setal," the effects of adjustment were felt in the degradation of urban moral and material "hygiene" in Dakar.[18] Three ethnographies of health in Senegal describe experiences of, and responses to, abandonment by the state. In the

1980s, Didier Fassin described the emergence of a new urban pharmaceutical political economy. With medicines lacking in public health care facilities, those no longer supported by a declining peanut economy found a thriving market for informal sales of medicines in the city. The state, which was unable to provide medicines but was also bound by patronage relations with the Muridiyya, a powerful religious organization that protected the illegal sellers (and which had shifted its economic base from the declining peanut to the emerging informal urban economies), could do little but stage the occasional and temporary shutdown of well-known illegal drug markets.[19] Growth in the number of (legal) urban private pharmacies also began in the 1980s and accelerated in the 1990s.[20]

Dovetailing these forms of privatization of health as side effects of the shrinking reach of the state, health care reforms in the 1990s officially made state facilities points of sale for medicines and care.[21] Ellen Foley has described the effects of a policy of decentralization, which devolved responsibility for health financing to local authorities, in a Senegalese medical district in the late 1990s. She observed that a rhetoric of "community participation" translated, on the ground, into sharpened disparities between different levels of ability to pay for care (including medicines, now being sold by public facilities following the implementation of the Bamako Initiative). This increased the vulnerability of the poorest as well as the burden on health care workers having to do more with less.[22] Duana Fullwiley's ethnography of sickle cell anemia in early 2000s Dakar also describes how patients were made to bear responsibility for alleviating their own suffering, and how they described their illness as an embodiment of economic crisis and governmental neglect.[23] This case also points to the emergence of an uneven cartography of public health following the increase of global health funding from the late 1990s. Well-resourced nodes have formed at the intersection of national public health and transnational resources (mainly for infectious disease, especially HIV), leaving large areas, especially of chronic, noninfectious forms of ill-health, "doubly abandoned."[24]

The waning ability of the state to pay for health has also affected health care workers. The anthropological literature has described a general shift from a bureaucratic, public logic of health care work to an increasingly personalized and commercial one.[25] Public health care workers in Senegal and elsewhere have been observed giving better care to those with personal connections or the ability to pay, selling medicines or services on the side (informally, or as part-time private practice), or seizing opportunities such as foreign-funded training seminars while neglecting the facilities under their responsibility.[26] These professionals have been depicted as agents of a predatory state, as indifferent, impolite, abusive, and even violent. Yet others have instead observed health workers' efforts to "improvise" care with limited means and technologies, upholding public service as an ideal despite shrinking budgets while keeping a critical eye on the distribution of access to care and the structural causes of ill-health. In these and other ways, a long-standing and deeply anchored ethos of public and national service through health work has been remembered and maintained.[27]

African scientists, the majority of whom are state employed, have also faced deteriorating work conditions since the late 1980s. Roland Waast and Jacques Gaillard described how African scientists were able to remain scientists only by adopting a new market-oriented ethos. Earlier, a conception of science as a public good, and of the scientist as a public servant, had been consolidated in the push by many African governments to construct and coordinate national systems of scientific research in the 1970s and early 1980s. Relatively well-funded foreign programs of scientific and technical cooperation delayed the impact of structural adjustment for a few years, but at the expense of national autonomy.[28] Many bilateral scientific cooperation programs were then drastically cut in the early 1990s.[29] Reduced state spending on universities meant that academic researchers (a high proportion of national researchers across Africa, just over 50 percent in Senegal in 1981)[30] faced rising student numbers with declining or stagnant operational budgets. Buildings deteriorated and laboratory equipment and supplies became difficult to obtain. Soon, public spending served only to pay salaries off which scientists could no longer live.[31]

To describe how scientists reacted to these changing conditions, Waast and Gaillard coordinated a team of about twenty researchers across fifteen African countries to undertake a quantitative study of scientific publications, local studies of scientists and institutions, and a broader questionnaire-survey of scientists. They found that, to avoid the same fate as public institutions that were turning into "shells of their former selves,"[32] scientists broke their allegiance with these institutions and took their services "outside the[ir] walls."[33] They took up consultancy contracts, sometimes from "fragments of institutions" such as labs or departments with good private or foreign connections, but more often by forming their own local NGOs and consultancy bureaus, usually on a project basis with international governmental and nongovernmental organizations, and with a short-term applied orientation. Thus an atomized and entrepreneurial scientific activity emerged, sustaining professional lives and identities despite institutional decay and fragmentation. In the space of a decade, the retreat of the state thus provoked a shift from "national sciences to a free market of scientific work."[34] While some African scientists and institutions have reinvented themselves with extraordinary success in this new market,[35] the shift has not come without tensions or a sense of loss. Waast and Gaillard state (perhaps overstate): "the anarchy of the market satisfies no one." They note concerns about the sustainability of "scientific capital" without investments in institutions and training, the lack of basic research, the effects of temporal fragmentation on scientific knowledge, and the critical stance of "researchers attached to their old national ethos."[36]

When the state no longer provides its agents with the means to combine professional survival with "being good" (as defined by ideals of public service), how can public health care workers and scientists preserve their material and moral integrity? Opportunism, disaffection, improvisation, and entrepreneurship, as well as nostalgia and aspiration, appear as different responses to this dilemma. They are not necessarily incompatible, and the choice or combination among them has probably depended on specific circumstances. These may give rise to hybrid institutional forms combining public and private, national, and transnational drivers and resources.[37] Many scientists have also developed versatile identities, as they worry about their own livelihoods and success but also about the public good. In the toxicology lab, institutional continuity was sustained in part by a work ethic anchored in both keeping busy and protecting publics, and yet, in the absence of stable funding, it also depended on "serving" international agendas and private clients, as well as personal ambitions and connections.

PROJECT TIME

In 2010–2011, lab members often emphasized, as they complained about things not working in the lab, that they used to have equipment that worked. An AAS (capable of detecting traces of heavy metals) now lay in pieces, they told me, in "Sokhna's room," the office of a junior analytical chemist.[38] A GC was still out on the bench, but in 2003 its printer began to malfunction; a Belgian expert was supposed to come but never did. "We can do a lot of things with a GC," one of the younger staff members explained, "amino acid profiles, pesticides. . ."[39] With either the GC or the

HPLC, it is possible to measure traces of the main classes of chemical pesticides: organochlorine, organophosphate, and carbamate. "[With] projects, there was equipment," Mamadou Fall told me, "since then there's nothing."[40] This suggests that the inactivated machines now stand as reminders of a time of temporary, foreign, and now-ended investments in the lab's capacity that served well-defined objectives. Indeed, Ba and Niane described projects as a transaction: the provision of capacity against the performance of specific tasks.

Projects, however, could also be animated by longer temporal horizons, as can be seen by taking a closer look at WACAF/2, the joint UN agencies project on marine pollution. Its objective was to provide the Dakar toxicology lab and other labs in West Africa with capacity that would be picked up, after the end of the project, by state monitoring mechanisms. The few clues I have about the "Italian project" indicate that it was a university-to-university partnership to provide equipment and training for pesticide residue testing in food (Niane's CV records that he spent two months in Sienna to train in methods of extraction and purification of pesticides for the measurement of residues in fruits and vegetables in 1990 and about three weeks in Montpellier to learn methods of pesticide analysis by gas chromatography in 1992).[41] If this is the case, this project, like WACAF/2, had capacity-building as its core objective, aiming to set up a durable infrastructure for regulatory testing (in this case, for food safety). Chapter 4 describes how Project Locustox was oriented by an even broader and longer-term ambition to create capacity to assess and monitor the large-scale effects of pesticides in West African climates and ecosystems.

WACAF/2 was one of a series of specific programs (WACAF/1–6) that aimed to provide the signatory states to the Abidjan Convention "for cooperation in the protection and development of marine and coastal environmental of the West and Central African region" with the technical capacity to carry out its Action Plan.[42] The convention entailed monitoring of marine contamination by pesticides, hydrocarbons, microbes, tar, petrol, and heavy metals.[43] UNEP was the main driving force for both the Abidjan Convention and the WACAF projects, which were part of its worldwide Regional Seas Program.[44] The organization provided about five-sixths of the total costs (632,700 USD) of the WACAF/2 pilot phase (1982–1984), which was combined with technical and financial support from the FAO, the WHO, and the International Atomic Energy Agency (IAEA).[45] From these resources, the lab was provided with the material required to measure heavy-metal traces in fish and seafood (the AAS and complementary equipment and chemicals) as well as guidelines, training, and intercalibration exercises to ensure the apparatus was used properly.[46] The lab's metal-measuring capacity was clearly tied to international resources, standards, and agendas for environmental and health protection.

Yet technical and financial responsibility for pollution control and monitoring was to be transferred to the convention's signatory states, which had pledged contributions to a regional trust fund. In other words, this analytical apparatus was to be durably maintained as a *regulatory* infrastructure by national governments. This did not happen. By 1989, states had still failed to meet their pledges, eliciting pessimism about the future of the Abidjan Convention.[47] The only work performed in the Dakar lab cited in a 1989 summary of WACAF project results had already been reported in 1985: fifty-six analyses performed over a period of six months, on six trace metals in six species of fish and seafood.[48] The 1985 report stated that regular testing of fish collected by the national oceanographic institute was expected to continue on a monthly basis, for an indefinite period, to watch closely for any rise in metal contamination.[49] Ba, then head of the lab, told me the project "lasted longer because it was financed by the FAO and WHO," so it is possible that fish testing continued after 1985 but was not recorded. However, it clearly did not, as planned, extend into a stable, locally funded future.[50]

Ba's statement portrays the monitoring of marine pollution as a transnational initiative, divorced from national priorities and limited to a well-defined (if long) project time. Yet the very nature and rhythm of the tasks carried out within the project, that is, the routine testing of parameters of contamination, bled into a more extensive temporality of regular monitoring. A narrowly defined project time, the time of funding and training, thus opened onto a (hoped-for) time of regularly repeated analytical activity that might attain governmental scales, on African national or regional levels, of space and time. This particular state-funded future did not materialize. However, the continued use of project machines by lab members did keep open the possibility of regulatory rhythms of analytical testing.

This has allowed lab members to remember project time not as a bounded time with precisely and externally defined futures but as open-ended, with fuzzy edges that bled into subsequent uses of the analytical equipment. Niane, Ciss, and Ba, for example, all spoke of heavy-metal testing in marine organisms as driven by the interests and investments of the

"FAO," but also by those of government, public authorities, and the export economy, blending the specific tasks of the project into the lab's orientation toward national economic and public health problems of potential contamination.

CONTRACT TIME

Lab members also associate the time of machines with the now-ended possibility of fulfilling public and private contracts, which provided an additional source of activity, and sometimes revenue, for the lab. By 2010, lab members were discouraged by their inability to satisfy requests from individuals, or even the police, for help finding traces of poison or contaminants. One day in the summer, two men came into the lab clutching a bottle of water that had suddenly flooded their house from an unknown source. All Professor Fall could offer them was to send the samples to France, at their own expense. He told me that the lab had been forced to warn the police and state prosecutors that, without money for equipment and supplies, the lab could no longer provide expertise in cases of suspected criminal poisoning or contamination. For Ba, such requests in the past evoked a sense of pride and duty, even though lab members sometimes "had to use their own means to respond," because "when the police ask you have to respond."[51]

I found records pertaining to forty-two analyses performed in the lab for "clients"—both public and private—between 1987 and 2008. These were in a file kept apart from the lab's main administrative records. The set is likely incomplete, as it contained no record of some of the contracts described to me by lab members (such as the analysis of pesticide residues on green beans for an agro-industrial export firm or the biomonitoring of employees of a local pesticide mixing and packaging factory). The file nevertheless gives a sense of the range of tests performed by the lab in response to external requests, and how these fit into—and might be remembered as part of—its activities as a whole. Twenty-one records were of requests to the lab by government services that included the national police, the customs agency, the civil aviation agency, and the Directorate of Commerce. The majority of these were for the identification of substances suspected to contain narcotics or the analysis of cadaver organs for traces of alcohol, narcotics, or poison (e.g., following a suspect death or an airplane crash). It is not specified whether services for the police, customs, or judiciary services were paid to the lab; they likely were not. Services to the aviation agency and to government control labs, such as the commercial control lab of the Directorate of Commerce, however, were billed. The latter, like the public supply pharmacy, requested tests of conformity (of food and drugs) that they should have been doing themselves, that is, according to their mandate, but were not equipped to perform. From 1997, the number of contracts with private companies increased. Of the forty-two records, seventeen were of requests by private and some partly privatized public companies for analyses of conformity (e.g., with displayed nutritional information in infant formula, or acetic acid concentration in mustard) and contamination (e.g., of pesticides and hydrocarbons in water).

These contracts increased the amount of benchwork in the lab, especially in the later 1990s and early 2000s, by paying directly for reagents (billed to clients) but also by generating extra revenue the lab could use to purchase supplies. A sense of continuity arose not only from the rhythms of this work but also from its content. Tests performed for public and private clients, as well as for transnational projects (with, as we have seen, public and regulatory aims) or for faculty and student research (who usually, as we will see, stated aims of improving public protection), were sometimes remarkably similar. Take the analysis of milk, for example. In 2002, samples of powdered milk were analyzed for bacteriological contamination as part of two different contracts, one for the government's commercial control lab and the other for a private import-export company. The same year, a thesis on the quality control of commercial powdered milks reported on the visual inspection and simple tests of solubility and fat and nitrogen content on milk obtained in Dakar's central market (Sandaga market), recommending improved monitoring and increased resources for government control labs (thus implying that their current capacity was weak or nonexistent).[52] A 1997 thesis reported a nearly identical investigation and conclusions.[53] That year, the lab also fulfilled a contract for the analysis of baby formula with a private company. A 1995 thesis reported results of HPLC testing of vitamin D content in powdered milk collected in Dakar shops, to determine conformity with labeling, also recommending the strengthening of the national lab responsible for fraud control.[54] Another thesis defended in 2001 reported the results of tests for quality and fraud of fresh milk sold in Dakar markets, recommending further testing for toxic contaminants.[55] Thus, the content of privately contracted work addressing the quality of milk as a commercial value bled into work performed for government control labs and thesis research initiated by lab members (leading to recommendations for improved regulatory monitoring) in which the purity, quality, and conformity of milk became an issue of public health and consumer protection.

Other products and types of analyses similarly crossed over between public and private initiatives, institutions, and interests, blurring the line between contract work, research, and regulation. While most analyses of "suspect powders" for narcotics were performed for the police or customs agency, one import-export company also requested drug and toxicity testing for soft drinks and candy in 1995. The lab tested drinking water for pesticide residues, first for the national water board in 1989, and then for its privatized commercial arm (created in 1996) in 1997. The lab also fulfilled contracts with private producers and distributors of pesticides for the biomonitoring of worker exposure and for the determination of active ingredient concentration. Several staff and student research projects surveyed pesticide residue levels in produce for local consumption, as an issue of public protection, but the remembered contract for green bean testing had involved commercial control of residues to meet export norms. Thus, in fulfilling contracts, the lab addressed issues of food safety, water quality, and worker health both as problems of private business management and as problems of public protection. Understandably, then, memories of fulfilling contracts are not of bounded bits of targeted and privatized analytical activity in the lab. Rather, they are of a combination of dutiful service to government and the public with the opportunity to generate some extra revenue from fees.

Project machines were used to fulfill some of these contracts. The requests and invoices I found did not always describe which analytical methods were used. Many, it seems, could be performed with relatively low-tech, largely manual methods such as thin-layer chromatography that did not require much equipment but did need consumables such as reagents (which cost the client from 10 to 200 percent of the price of analytical work), glassware, and special filter and chromatography paper. Yet the presence of the machines in the lab, given that few other labs in the country had this equipment, may well have attracted important contracts for the detection of contamination by pesticides, heavy metals, and hydrocarbons, such as the long-term, biannual monitoring of Dakar's drinking water reservoir. Lab members speak vaguely of a time "of many contracts" as a time when the machines worked, and when it was possible to purchase renewable lab supplies, whose versatile capacity filled the lab with many different types of activity.

ROUTINE RESEARCH

"It's research that can solve problems. It's not so cutting-edge, but it's useful." Bara Ndiaye, an analytical chemist, described some of the work conducted in the lab for student theses and staff publications in the 1990s and early 2000s. These included analyses, for example, of the quality and composition of butter, powdered milk, edible oils, and generic drugs. This struck me as the kind of work that should have been conducted by government control labs. Ndiaye agreed.[56] What were the problems this research "solved": unanswered questions about the quality and safety of food and drugs, or the fact that regulatory laboratories were not functional? It is because these labs could not fulfill their responsibilities for monitoring foods, drugs and environments by performing regular tests, that lab members and students could perform such analyses under the label of research.[57]

Earlier claims to original research by lab members in the 1960s and 1970s were justified by methodological innovation. From the mid-1980s, however, most of the research conducted in the lab was defined as original not by its methods—which were routine—but by its samples, as otherwise untested, unmonitored matter.[58] Niane remembers this period, vividly, for the grueling task of collecting samples "under the hot sun," in fields, and in markets, and for putting in time at the machines rather than for modifying methods.[59] In order to get research done, lab members and especially their students set out into the largely unexplored (by chemical analysis) terrain of the Senegalese economy, its fields and markets, and brought back new things to put into the machines.

Occasional mentions and diagrams of the AAS, GC, and HPLC appear in the pages of student theses stuffed into metal filing cabinets and piled on lab members' shelves at the university. During and after WACAF/2, students collected additional samples of fish and seafood to analyze for trace metals; they also inserted samples of green tea (widely consumed in Senegal) through the AAS to detect traces of fluoride, lead, and copper, and tested tree bark to measure lead residues from urban gas emissions.[60] The HPLC was used in a series of student projects defended between 1997 and 2001 on levels of aflatoxin in artisanal edible oils, particularly peanut oil and paste, collected from markets in various regions of Senegal.[61] One thesis situated itself as part of a series of studies in the university lab "on the theme of aflatoxins and food," driven by a "sustained and constant search of improved nutritional quality and atoxicity in food products among others, to be made available to the population."[62] The GC was also used in analyses of the quality of oils and butter, as well as to search for pesticide residues in well water collected in a rural area (where a pediatric research unit had a field station) and in breast milk (a commonly used indicator of a population's exposure to persistent pesticides).[63] Niane was among the students who benefitted from the presence of machines to complete his thesis work. He used the AAS for his graduating pharmacy thesis on heavy metals in fish and seafood (work he conducted from 1987 to 1990, thus overlapping with the WACAF project and similar in nature).[64] Later, he used the GC to conduct research on pesticide residues in peri-urban agriculture for an advanced studies diploma (*diplôme d'études avancées*, or DEA) in biochemistry completed in 1993.[65]

This research allowed Niane to stay and rise in the lab's hierarchy. With the first degree, he moved from technician to teaching assistant, which allowed him to supervise practical exercises. The second degree allowed him to take on lecturing duties.[66] Other students' thesis results sometimes led to publications coauthored with staff members, while the latter conducted additional research using this equipment, for example, on the use of plants as indicators of pesticide residues in the atmosphere and the food chain. Most of this Dakar-based work, however, was only published locally, in the medical journal *Dakar Médical*. Overseas research, generally doctoral research, accounted for most publications obtained in French and European journals during this period, in contrast with earlier times, when Gras and his colleagues had published their local work in such journals (see chapter 2).[67] Presumably only the best (or favorite) pharmacy students were given access to project machines and lab supplies for their thesis work; the majority had to make do with bibliographic essays or compilations of clinical data and questionnaire results. Yet with few or no jobs opening up in the public sector, and with government labs understaffed and underequipped, while private clinics, labs, and pharmacies were proliferating, most graduates did not pursue careers involving toxicology-related research or regulatory testing.[68]

For the duration of their thesis work, however, in their final year before graduation, a few dozen students briefly performed the kinds of regulatory testing they, with their supervisors, called on the state to conduct routinely.

Thesis students and sometimes junior lab members, rather than government agents,[69] went out to collect fish, vegetables, tea, coffee, milk, peanut butter, or edible oils to measure for conformity and contamination. Sampling at points of sale (in shops and street stalls) is particularly significant, since it captures products to which the Senegalese population is actually exposed to, as opposed to testing done on products entering or leaving the country. It thus put strategic locations onto a patchy map of food- and drug-quality regulation that, even as it has improved in the 2000s, has tended to prioritize the control of traffic into and especially *out* of the country, into better regulated export markets. Thesis work also introduces a preventive temporality of exposure detection, while official responses—as can be seen in police and judiciary requests for testing—have been largely limited to acute poisoning, often as a suspected cause of death.

Despite its modest objectives, this research conducted by thesis students and lab members represents brief instances of what operational public health and environmental regulation of toxic contamination might be. In their more ambitious recommendations (generally for better regulation), research articles and especially student theses can also be seen as sites of regulatory imagination, projecting the routine (but abbreviated) rhythms of regulatory testing actually performed by their authors toward ideal futures of continuous, regular, and durable monitoring of toxic risk. In a typical example, one student who tested artisanal peanut oils for contamination by aflatoxins, which the Senegalese government was tackling primarily in order to meet export norms, recommended "the establishment of a *real* food surveillance policy including *regular* and *systematic* controls."[70]

What Ndiaye alluded to as routine research, research that "is not cutting edge" yet "useful," thus activated several temporalities. This research kept project machines in action beyond the end of projects, but not exactly as envisioned in project objectives. The WACAF projects had planned for continued monitoring of marine pollution to be sustained by African state financing. Instead, scientists drew together the lab's meager operational budget and revenue from analytical contracts (and probably contributions from thesis students' bursaries) to support modest research projects. Defined by lab members rather than by national or international policymakers, these momentarily probed the Senegalese environment for toxic (and other) risks. This research advanced lab members' careers, albeit more slowly than the more cutting-edge graduate or visiting research they conducted overseas.

Much of the productivity of machines was fueled by the work of junior lab members, notably Niane as well as thesis students. In our conversations, Ciss was very emphatic about the key role Niane played in making the machines functional. Referring to himself and Ba, Ciss explained that, with teaching and administrative tasks: "We don't have time! But with someone who is there like Babacar Niane . . . it's relatively easy. We give him work and we are sure he is going to do it." And later: "We had freezers, the GC, the HPLC . . . and also an AAS, there was equipment. And it worked. And also, there was Niane. When there was a problem . . . so he played a key role! Oh yes, yes, in the life of the laboratory."[71] An essential actor in fulfilling the tasks assigned to the lab by the projects, Niane was able to raise his own status using the analytical capacity they left behind. Yet even before the GC broke down, but after salaries were devaluated and money for lab consumables ran out, Niane felt he was forced to leave, interrupting his forward-moving trajectory through the lab to secure his livelihood in commercial pharmacy. And for most of the students who conducted theses in the lab, this time of research was their only opportunity to be "civic scientists," performing routine regulatory testing briefly, activating a sliver of surveillance over a vast Senegalese landscape of potential contamination.

THE REGULATORY IMAGINATION

Recently retired from the university, after heading the lab for more than two decades, Ba summed up its activity for me as pertaining to three axes: heavy metals, pesticides, and aflatoxins. In describing these axes, Ba suggested the lab had had a coherent orientation, largely defined by its own members, which targeted the principal toxins that threatened Senegalese public health and its export economy. In these three categories of toxic matter, past analytical capacity, especially (but not exclusively) the capacity provided by the machines, drew together the objectives of transnational projects, private contracts, and public requests for expertise, and those of modest research projects initiated within the lab. For Ba, the lab had clearly had a research *program*, implying a degree of coherence, autonomy, and continuity, generating memories of work he described as "relevant to Senegalese priorities,"[72] or, as Niane said, "of interest to government . . . to public authorities," made up of research that, in Ndiaye's words, "solved problems." This, according to Fall, is no longer the case: "We don't have a research program, what, with our budget. It's à la carte; when there are means, when there are opportunities."[73] Without continuous capacity, research splinters into projects. This results in a sense of loss of autonomy but also of a sense of "being good" by serving the public.

Yet the continuity of capacity and activity in the lab from the mid-1980s to the early 2000s, as remembered, somewhat nostalgically, in 2010, did little at the time to alleviate discontinuities in national circuits of environmental and health protection. On the contrary, the very moments remembered and recorded in the lab lay bare the patchiness of toxicological testing. These are brief moments of testing that *resemble* regulatory monitoring, which were performed as part of time-limited student or international projects as well as contracts (few of which were concerned with preventing toxic exposures). They did not add up to any continuous or comprehensive coverage on a national scale. They predict poor likelihood that molecules of pesticides clinging to vegetables headed to local markets or leaching into well water, or of carcinogenic toxins released by fungi growing in artisanal peanut butter or oil, would ever meet a reagent and become an analytical result or protective action. For these molecules and the populations within whose bodies they might accumulate, the state as regulator of the circulation of poisons manifested as "'absent' presence," as has been observed of the African state under adjustment in other areas of social provision.[74] And still, even while recognizing the limited reach of their activities, lab members speak of a time of regular and "public interest" work.

What toxicological testing in the lab generated was not effective regulation but what we might call a regulatory *fiction*. As fiction, the governmental control of food, drugs, and the environment during this period, like deteriorating health care systems and increasing poverty, contributed to the persistent or growing "unprotection" of Africans that has been documented by medical anthropologists.[75] As with the introduction of user fees for health care or lack of control over the rising cost of basic foodstuffs, patchy toxicological testing is a failure of the state to moderate the exposure of its citizens to the circulations (including of toxins) and exclusions of the market. A dramatic illustration of unequal protection to toxic exposure in domestic and export economies can be found in the explosion of an ammonia tank in a peanut-processing factory in 1992. Ammonia was used in a process for detoxifying (from aflatoxin) peanut cake, a by-product of oil production used as animal feed, in order to meet European quality standards for export.[76] Peanut production was declining but still accounted for a large percentage of Senegalese exports. Meanwhile, 129 workers were killed (in part due to the absence of emergency medical services and hospital understaffing) and many more harmed by the explosion, while many poor Senegalese continued to eat contaminated peanut products.[77] One could tell similar, if less dramatic, stories about the testing of pesticide residues on produce destined for local and export markets.[78]

Yet as a fiction, toxicological regulation also opens up possibilities for imagining—not just stating the absence of—an effectively protective and provisioning state. Is such imagination nothing more than pretense and performance? Scholars have noted how eroded and adjusted African states have not vanished but have continued to exert power either through highly visible but largely performative and/or episodic presence or through invisible parallel circuits.[79] Adeline Masquelier elegantly captured the paradoxes of the Nigerian state's "'absent' presence" in the image of a dispensary's "prosperous facade." Behind the freshly painted walls of a rural dispensary, she observed, was emptiness: the staff's inability, without adequate stocks of medicine or fuel for medical evacuations, to dispense care. Yet this emptiness hid another layer of privatized care animated by stocks of medicines made available to those with connections, or smuggled into the black market. Between these two layers of illusion, the dispensary continued to exert a power that was both coercive and disciplinary through the possibility of treatment, arbitrarily rationed, and through rumors of abuse. Masquelier writes, "It is the simultaneous absence and presence of prosperity and privilege that creates a space of desire filled with the collective fantasies of ordinary citizens who struggle to make sense of the paradox of a state that exists largely by virtue of its unreality. In this space, people can both decry scarcity and indulge in the pretense that the dispensary is full."[80]

Lab members' partial enactment of regulation—testing things for contamination without the repetition over space and time (or of action taken on the basis of findings of contamination) that would qualify it as effectively protective—can likewise be seen as an empty performance of public service held up by the pretense of a functional state. It should be noted that this "facade," unlike the dispensary's, was not a highly visible one to the Senegalese public. While the state made some promises of toxic regulation, for example, by maintaining ("empty") quality control labs or signing international conventions on marine pollution, it did so to restricted and often distant publics. In addition to the low priority assigned to the monitoring of contamination in African public health, the bodily imperceptibility of nonacute toxic exposure probably helped keep the promises and patchiness of toxic protection out of the public arena.[81] Yet in a way, this invisibility places an even greater responsibility on toxicologists (and analytical chemists) as those with the ability to make toxins, as well as potential gaps in their regulation, visible.

Thus I suggest that the regulatory imagination in the lab was also a critical and productive one. The fiction of regulation was not a delusion: during my fieldwork, I heard several lab members decry, informally as well as in public speeches, how poverty and lack of political will exposed Senegalese populations to contaminated food and poor-quality products (e.g., "the Senegalese have no idea of [what's in] what they eat!").[82] There is no reason they would have thought differently ten or twenty years earlier. On the contrary, in the 1980s and 1990s, they had not yet, as they had by the end of the 2000s, begun reanimating the ruins of public projects of food, drug, and poison control. Yet if testing in the lab did not equate to protection, it did monitor the toxicity of real bits of what Senegalese people were exposed to. The brief and partial, yet successive, enactments of a "good state" by university scientists pointed out particular problems of exposure in zones of regulatory absence (in particular, pesticide residues on vegetables and aflatoxins in artisanal peanut products) but also pointed *toward* the extension, through regular repetition and scaling up, of this testing as effective public protection. Thus remembering continuous activity in the lab as being in the public interest both, in a sense, *pretended* that it operated within an effectively protective state and *called upon* the state to *become*, by taking up responsibility for the extension of this activity, what these fragmentary actions revealed it was not.

4 · Prolonging Project Locustox, Infrastructuring Sahelian Ecotoxicology

Among the international projects that enlisted the university's toxicology and analytical chemistry lab, LCAT, was one that is commonly referred to (including in official documents) as "Project Locustox." It began as a three-month pilot project conducted in northern Senegal for evaluating the environmental effects of locust and grasshopper control. This FAO-sponsored initiative aimed, in particular, to perform such evaluation in the type of environment affected by large-scale pesticide spraying that targeted these insects' breeding grounds. During the pilot, lab members Babacar Niane and Mounirou Ciss were put in charge of collecting and preparing plant and animal samples in test fields, as well as performing some manual analyses (those requiring more sophisticated equipment were done in Europe). For Niane and Ciss, this was a brief collaboration that left few, indistinct memories of hard, hot fieldwork. It brought no major equipment or training opportunities to their lab at the university. The project itself, however, was carried on in a subsequent phase, followed by another, then another. With each prolongation of the project, ecotoxicology, a science combining toxicology and ecology, was tied increasingly tightly to West African ecosystems and institutions. By the late 1990s, foreign (mainly Dutch) scientists and their national (Senegalese) counterparts were searching for ways to convert this legacy of accumulated capacity into a permanent local institution. In 1999, CERES-Locustox/Centre régional de Recherches en Ecotoxicologie et Sécurité environnementale (Regional Center for Research on Ecotoxicology and Environmental Safety) was created as a private foundation with a public mission, with a three-year transition phase of continued support from the Dutch government and the FAO.

This chapter leaves the university (UCAD) to follow another group of scientists who, like the lab members of the previous chapters, worked toward regular, continuous, and cumulative rhythms of activity in order to create and maintain capacity for an African toxicology. Like the university lab, Locustox was a site for the pursuit of public and protective toxicological science that was firmly hinged to national infrastructures, yet which also relied heavily on additional, nonnational sources of support. The study of Project Locustox as it unfolded in the 1990s similarly illuminates efforts by scientists to hold up the production of knowledge as a public good during a time of cuts in funding, both national and transnational, to African science.[1] Unlike the lab, however, indeed unlike many other national institutions that survived thanks to transnational collaborations in Africa during this period, Locustox's scientists were able to maintain sufficiently high levels of funding for long enough to conduct cumulative research that generated novel and distinctively local scientific methods and knowledge.

In this chapter, I examine how this localness, which resided in the adaptation and integration of ecotoxicology's methods both to local ecosystems and to local regulatory institutional networks, played out in arguments for prolonging research and funding. The project's starting point was the recognition that ecotoxicological methods and standards had been created mainly in and for temperate climates and ecosystems, while locust control happened mainly in hot and dry environments, and at particularly large scales. Thus an ecological and epistemological rationale (different ecologies demanded distinctive practices of knowledge production) was mobilized to justify the time and money needed to accumulate detailed data on ecological populations and relations in West African environments. Yet the project was also closely tied to a more political argument to durably anchor this science in public West African research and regulatory institutions. Tracing how accumulation, localization, and regulation both warranted and resulted from prolonged support for ecotoxicological science, I take Project Locustox as an exceptional case study of relatively successful efforts to make science simultaneously durable and protective in Africa under structural adjustment.

When I conducted research in 2010, several scientists outside CERES-Locustox, especially at the university, spoke of the research center as a well-equipped but expensive lab (referring to its environmental chemistry lab) for pesticide residue analysis. Unfortunately but understandably, the center had gone commercial, charging high fees for its services. At the center itself, the activity of the environmental chemistry lab, kept busy by contracts from agro-industrial firms, contrasted with its much quieter vertebrate and invertebrate biology departments. Still, the past of Project Locustox, a past of studying ecological relations and measuring their disruption by exposure to pesticides, was still palpable in the presence of fish—used as indicator organisms for pesticide toxicity—reared in aquariums and concrete basins, old pickup trucks bearing FAO logos, nostalgic memories of counting dead insects in fields and ponds, and persistent hopes of reactivating capacity to conduct regulatory testing of environmental toxicity that only governmental institutions would want and be able to fund. This chapter is about the nature and creation of this legacy of African regulatory ecotoxicological science—that is, of how, and for what, it was built up and made to last. It is also about the challenges and contradictions of keeping this fragile legacy alive through the mobilization of entrepreneurial strategies.

LOCUSTS, VULNERABILITY, AND "INFRASTRUCTURABILITY" IN THE SAHEL

In the late 1980s, an invasion of desert locusts of unprecedented proportions prompted the spraying of pesticides over vast areas in their breeding grounds in the Sahel, the arid and semiarid band of land south of the Sahara. The scale of the invasion, and the magnitude of the response, raised concerns about the vulnerability of West African ecosystems and the fragility of its locust-monitoring and crop-protection infrastructures. A long period of latency since the previous locust upsurge, in addition to the impact of structural adjustment, had eroded the capacity of state-funded national and regional institutions to predict or mitigate such invasions. The FAO and the United States Agency for International Development (USAID) were the main providers of emergency financial and technical assistance for pesticide spraying in the Sahelian area.[2]

FIGURE 4.1. Fish, raised as indicator species, CERES-Locustox.
Photo by Noémi Tousignant, Dakar, 2010.

FIGURE 4.2. Fish tanks for raising indicator species (young), CERES-Locustox.
Photo by Noémi Tousignant, Dakar, 2010.

FIGURE 4.3. Fish tanks for raising indicator species (mature), CERES-Locustox. Photo by Noémi Tousignant, Dakar, 2010.

Both agencies had, in previous years, adopted more stringent environmental policies and were accountable to publics outside Africa that were growing increasingly concerned with pesticide risks to human health and ecosystem equilibrium.[3] Public attention to the unequal geography of these risks in a "global pesticide complex" also grew in the 1980s.[4] This concern was informed by shocking and accessible accounts of the "circle of poison" (the export of banned pesticides from wealthy to poor countries and their return as residues on imported produce),[5] of "Bhopal syndrome" (the relocation of pesticide plants to less regulated industrial environments),[6] and of the amplification of pesticide risks by irresponsible marketing, weak legislation, and poor farmer education among the "Third World poor."[7]

The FAO and USAID took different approaches to locust control and its potential ecological risks. The USAID sponsored at least two environmental assessments. The first was of a specific USAID-funded spraying operation in Senegal in 1986. Its report concluded that the sparseness of life and water in the Sahel "somewhat reduced the chance for a significant environmental impact," with populations of potentially vulnerable nontarget invertebrates (species that are unintentionally affected by pesticides with adverse ecological consequences) "so low that no impact was observed."[8] The second, published in 1989, addressed control operations across Asia and Africa. It relied on data obtained in temperate environments and organisms to rank pesticide toxicity. This was not seen as a problem: in any case, better knowledge of geographically specific pesticide ecotoxicity would be of little use to evaluating the control strategies recommended to USAID. The 1989 report (which noted the weakness of West African national and regional pest control institutions) advised to invest in locust alert systems "at the village level," through the recruitment of volunteers and farmer education, which would help to minimize pesticide use.[9] A later report was even more explicit in its advice to bypass government channels via "decentralization, such as promoting village brigades, local control and increased investment by non-governmental organizations and the private sector."[10] In other words, USAID consultancy reports outlined a strategy to address locust and grasshopper control (to protect food security, and therefore political stability) and its potential risks to the environment that would not require investments in African science, or the building up of African governmental research and regulatory capacity.

By contrast, the scientists who advised the FAO's Desert Locust Control Committee insisted on the need for specifically Sahelian toxicity data to assess and monitor locust control. They also emphasized the potential for Sahelian biological organisms as well as government institutions to become infrastructure for ecotoxicology. In other words, they bet on the "infrastructurability" of the Sahel for generating a distinctive form of ecotoxicological science. It was at a meeting on the committee's research priorities held in 1988 that Ian Grant, an ecologist at the British Overseas Development Natural Resources Institute (NRI), argued for such a Sahelian ecotoxicology. Grant stated that temperate data and methods were unsuited to the environmental study of locust control and outlined a program for selecting "sensitive ecosystem components" and obtaining data on processes of "natural variation," which would require extensive ecological groundwork.[11]

Support was thus secured from the Dutch, British, American, French, and Senegalese governments for a three-month pilot study in northern Senegal to "determine a set of taxa and/or ecological processes to be studied in greater detail."[12] The study team was made up of four Dutch ecotoxicologists; a locust and grasshopper specialist, and an ecologist from a French agricultural research center; an ichthyologist (fish scientist) and an ecologist from the United Kingdom; an American ecotoxicologist/ornithologist; and, from Senegal, two university toxicologists (Ciss and Niane) as well as a termite specialist from the Forestry Services and an entomologist from the Plant Protection Services (Direction de la protection des végétaux, or DPV).[13] Together, they surveyed the aftermath of experimental aerial applications and "over-applications" of three chemicals commonly used in locust control (chlorpyrifos and fenitrothion, both organophosphates, and the insect growth regulator diflubenzuron) in large plots of dry savannah (2 × 2 km and 2 × 3 km) and in artificial lakes. Some watched birds; others trapped grasshoppers. The corpses of animals and insects were collected and counted. Ciss and Niane cut grasses at precise intervals of time and space, which they homogenized, weighed, and wrapped for freezing and shipping to the Netherlands for the measurement of pesticide residues. They also brought back bird brains, grasshoppers, scorpions, fish, and shrimp to the university lab in Dakar[14] to test for levels of cholinesterase—an enzyme inhibited by, and therefore indicative of, exposure to organophosphates.[15]

The pilot study was headed by James Everts, a toxicologist at Wageningen University in the Netherlands, working as a consultant to the FAO. Everts's final report of the study was a plea for ongoing support in order to build on the work initiated during the pilot phase. Like Grant before him, Everts questioned the applicability of available knowledge about pesticide toxicity, 90 percent of which had been produced in temperate environments. Everts also criticized previous studies of the environmental effects of locust control (in which he likely included the 1986 assessment sponsored by USAID) for their "fractional, local and temporary, short-term character."[16] By "local," he meant small, for he also accused them of having an insufficient grasp of local ecological dynamics; they were hampered by a "vague conception of the indicator parameters to be monitored and their relevance to the ecosystems concerned."[17] Bigger, longer studies were needed to determine the specific vulnerability of Sahelian life, which Everts characterized as concentrated rather than sparse, gathering around "biologically rich" areas to form "a limited number of processes that are sensitive to disturbance by pesticides."[18] While the pilot study had already begun the work of identifying good indicator species, it had cast its net wide to count organisms of numerous species found dead or alive in sprayed plots. Refining this selection, Everts explained, required highly detailed knowledge of species' sensitivity to changing environmental conditions (in this case, to pesticide exposure) and of the significance of their role in local ecosystems. Further data was also needed on the "background noise" of constant ecosystem change to set against the potentially disrupting effects of experimental spraying. Everts thus described making indicators as long and painstaking work, requiring sustained ecological observation and repeated manipulations in order to specify species' survival, abundance, mortality, adaptability, reproduction, life cycle, and so on.

Everts's recommendations did not just call for an increase in the scale of subsequent studies. They outlined a plan for moving deeper and more durably into the Sahel as a point of contact between ecosystems, pesticides, and scientists. This entailed a dual commitment to place-specific methods and to the building up of existing national governmental scientific capacity through "enhance[d] collaboration and training."[19] Indeed, as Everts explained in his report, the location of the pilot study had been selected in part due to "existing links" between Wageningen University and "major national services" such as Senegal's DPV and national agricultural research institution (Institut Sénégalais de Recherche Agronomique, or ISRA), as well as the Dakar-based headquarters of OCLALAV, the Organisation commune de lutte antiacridienne et antiaviaire, a regional (interstate) organization for the control of locust, grasshopper, and bird pests.[20] More generally, Everts justified the choice by referring to the area's "good infrastructure" and "relatively well-known ecology," as well as skilled local personnel.[21] Thus Everts offered as resources for Project Locustox the products of past public African science: experienced government scientists, knowledge of local ecologies, and the (residual) capacity of national and regional institutions. Everts's vision of the richness and vulnerability of Sahelian life may indeed have been based on the work of Senegalese scientists such as Vincent Dieme, a DPV entomologist who had previously worked on indicator species and would play an important role in Project Locustox.

While the USAID-sponsored reports had characterized the Sahel as inhabited by sparse life and dysfunctional government institutions, Grant and especially Everts depicted it for members of the FAO's Desert Locust Control Committee as a space inhabited by fragile but potentially valuable organisms and organizations. This vision underpinned a commitment to extensive indicating work, on and with Sahelian insects, in spaces shared with governmental institutions. In Everts's 1990 report, the *Sahelization* of ecotoxicology entailed its penetration of the local both as a natural environment—a set of ecosystem dynamics that developed under specific patterns of heat and humidity—and as a network of existing governmental institutions. Prolonging Project Locustox was both an epistemological project and a political one, seeking to align the science of ecotoxicology with the natural and institutional environments of the Sahel and to generate both Sahelian indicating methods and Sahelian/Senegalese public science.

PHASE TWO (1991–1994): ECOLOGICAL CONNECTIONS, INSTITUTIONAL CONFUSION

Apparently, Everts's arguments were persuasive. With Dutch and Senegalese government funding, the FAO launched a second phase of Project Locustox in 1991. The work of selecting suitable indicator species began. As scientists followed and connected Sahelian life-forms in order to identify nontarget organisms and describe their ecological roles, they also wove together the project's actions and people—foreign experts paired with "national counterparts" and a string of students and interns from Wageningen and from West African governmental agricultural and crop-protection training programs—with national research infrastructures.

Studying the breeding grounds of "target pests"—locusts and grasshoppers—project scientists observed the beetle *Pimelia senegalensis* feeding on the young, vulnerable egg-pods of locusts. Two adjacent ecosystems were also potentially vulnerable: the agro-ecosystem of millet, the crop most exposed to locust-control spraying, and the water ecosystem of temporary ponds that form during the rainy season. Millet's main pest is *Heliocheilus albipunctella* or the millet head miner. Its parasite is the wasp *Bracon hebetor say*, which feeds on the larvae of the millet head miner. The beetle *Pimelia* and wasp *Bracon* are therefore nontarget species. If weakened or killed by pesticides, crop pests such as locusts and millet head miners would be relieved of parasitism and allowed to reproduce more vigorously. Thus the crop protection effects of locust-control measures would be partially reversed. In temporary ponds, the scientists searched for species that were both abundant, suggesting an important ecological function, and sensitive to pesticides. They eventually selected a fairy shrimp, *Streptocephalus sudanicus*, and the water insect *Anisops sardeus*, a backswimmer.[22]

Some of the work toward locating nontarget species had been done previously by government scientists working with the DPV, OCLALAV, or CILSS, the Permanent Interstates Committee for Drought Control in the

Sahel.[23] For example, DPV entomologist Vincent Dieme had worked on the *Bracon* wasp as part of a prior CILSS-coordinated project.[24] Yet there was still much to be done toward developing field- and lab-based methods for manipulating these organisms for toxicity testing, and to collect the background data needed to interpret the significance of their response to pesticide exposure. This entailed the creation of study sites and the development of modes of manipulation and circulation between fields and field labs that would yield information about pesticides' environmental toxicity.

Locustox study sites, as well as its organizational chart, were grafted onto existing DPV structures. Describing one of the project's antennas in northern Senegal (where the pilot study had been located, near Richard-Toll), an intern wrote in his report that the overlap of the site with a DPV outpost, of which the director and staff had been named temporary local counterparts, created "institutional confusion." This was, he explained, "because there are DPV structures on which the project acts, to make them more functional and get them going, such as the entomology laboratories."[25] At the Richard-Toll site, *Pimelia* were trapped in experimental plots, then sorted and counted in the field lab. Attempts also began to raise them in the lab, but the beetles' long life cycle made wild organisms more convenient for lab testing. The captured organisms survived well in the lab or in insectariums; they required little maintenance and reproduced at high rates.[26]

This overlap of ecological connections and "institutional confusion" was even greater at a second project site, which was also located near a DPV field station at Nioro du Rip, in southwestern Senegal. This site was dedicated to the study of the millet agro-ecosystem in farmer-cultivated fields. For this study, project technicians partitioned the land in which local farmers grew millet into blocks of control and test fields. They then carefully sprayed the latter with equal measures of fenitrothion and diflubenzuron. Moving between the field and the DPV field lab, they were able to confirm that *Bracon* was both a parasite of the millet head miner and sensitive to pesticides, and thus a good candidate for indicating.[27] A doctoral student from Wageningen University, Jan van der Stoep, spent a six-month practical internship developing a rearing method to ensure *Bracon*'s availability for lab screening tests.[28] Van der Stoep built on Dieme's previous unfinished work on *Bracon* and owed much, he gratefully acknowledged, to the help of Baba Fall, the local DPV lab technician who cared for both the wasp colony and a host colony of *Ephestia kuehniella* that was reared in the lab as a food supply for the wasps. Fall helped to transfer *Ephestia*, a pest of stored millet, from nearby infested stocks. He also invigorated the colonies from time to time by bringing in wild wasps and fresh millet leaves from the field, and brought lab wasps out to conduct tests on them in the field, to "link the laboratory tests with the field situation."[29] The rapid adoption of the wasp as a lab organism was thus a collaborative achievement involving constant circulation between sites and practices of caring, feeding, growing, capture, and control.

It was also near Nioro du Rip, and its existing DPV facilities, that Locustox staff created a site to study the ecosystem of temporary ponds. For two weeks, they surveyed and selected sixteen ponds. These were monitored in 1991 to measure invertebrate population densities before and after experimental spraying, and in 1992, for recovery and ecological studies. They also placed shrimp in cages in the ponds for in situ tests of the death rate due to acute exposure to pesticides and sent water samples for purification, extraction, and analysis for residues in the central DPV lab near Dakar, as well as a GTZ (the German government technical cooperation agency) lab in Germany.[30]

PHASE THREE (1994–1997): OF DEATH AND REGULATION

By the end of the project's second phase, three candidate indicator species were identified: *Pimelia*, *Bracon*, and *Anisops*. More information was needed on the latter's "foodweb relationships, spatio-temporal distributions, reproduction, migration patterns and abundance," and their use in toxicity-testing methods was not yet established.[31] More progress had been made in turning the other two species into indicators: the beetle *Pimelia* was declared part of standard operating procedure for toxicity screening, and the rearing of *Bracon* was "mastered" in the Nioro du Rip field lab.[32] Still, the reports on phase two studies, and the report of a technical consultant from Wageningen, Frans Meerman, who visited Senegal to advise on the next phase of Locustox, called for more work to determine what, exactly, these indicators indicated. The larvae of the beetle *Pimelia* had been confirmed, through the monitoring of routine DPV spraying, to be the main predator of grasshopper egg-pods. Yet further information was needed on the beetle's natural enemies and its life history, on which no published data could be found, in order to distinguish between "background" patterns of survival and mortality, and the outcome of their disruption by pesticides.[33] The millet head miner was said to be in "urgent need" of a good lab diet. This would allow for incubation studies to determine just how much of its mortality was caused by the parasitoid wasp *Bracon*.[34] Studies of the range and distribution of the head miner's natural causes of death, as observed in the field, were also necessary to determine the wasp's exact rank among its natural enemies.

Meerman's 1993 report, which set out why a further project phase was necessary, explained that *Bracon* would have limited value as an indicator without such additional data on the millet head miner's mortality patterns. Results of tests using *Bracon* would be difficult to interpret: how exactly did they translate changes in complex and potential adaptable pest–pesticide–natural enemy interactions? Meerman called for the creation of life tables of the crop pest and of other pest populations, that is, a quantitative description of species' mortality patterns and factors. Because ecological data on most Sahelian species were unavailable, it would be "difficult to answer this question by a study that has a limited timeframe." Thus he called for the prolongation of ecological observation over several years. At the same time, he called for a clearer specification of institutional roles among Locustox and the DPV. This was needed to determine how the knowledge obtained from indicator species such as *Bracon* would be used to inform the policy and practices of the FAO and DPV (and presumably other potential users of its data, such as the CILSS).[35] In other words, clarification of both complex ecological interactions and the modalities of institutional interactions were needed to smooth the path from indicating tests to regulatory decisions.

During the project's third phase, observation thus continued on the life cycle and mortality patterns of the millet head miner. Two life tables were created by interns from Senegalese government–run agricultural training schools: the Ecole Nationale des Cadres Ruraux (National School for Rural Officials) in Bambey and the Ecole Nationale Supérieure d'Agriculture (National Advanced Institute of Agriculture) in Thies. The interns worked during 1994 and 1995 in a millet plot maintained by ISRA, the agricultural research institution, in Nioro du Rip.[36] Listed as first authors of the FAO-published reports, the interns explained how this work guided the selection and use of natural enemies as indicators, as well as the development of pest-management strategies.

Two additional years of temporary pond observation and intervention confirmed the fairy shrimp's and backswimmers' ecological roles, and their utility as vulnerable experimental organisms.[37] Tests of other potential indicators also increased the density of institutional and collaborative relations around the project, for example, the use of fish as indicators, which were taken for toxicity testing to an ISRA lab in Fanaye, and observations on other kinds of beetles, of which the identification was confirmed at the IFAN (Institut Fondamental d'Afrique Noire) in Dakar.[38]

As evaluation methods were standardized, the observed effects of experimental treatments, which had previously fed back into methodological development, could now be converted into regulatory knowledge and education. By 1996, the methods developed during the second and third phases of the project generated much of the data on desert locust and grasshopper control that served to rank pesticide environmental toxicity for the FAO's Pesticide Referee Group.[39] The project also began monitoring the exposure of DPV pesticide applicators. In 1995, a "training and information" component was added to the project, seeking to diffuse project results to "potential partners" such as the staff of the DPV and the functionaries involved in crop protection and pesticide registration at the CILSS. Training included the acquisition of new skills by project staff, as well as the diffusion of basic information about pests and pesticide uses across national space, including at the level of village control committees.

AFTER THE PROJECT? BUYING LEGACY

The third project phase also initiated a conversion of Locustox from project to national institution. From 632,000 USD in phase two, Dutch government funding jumped to 3,421,031 USD in phase three to build up a transferable legacy.[40] Training was an important component of this planned legacy, but material capacity was also central: new buildings were built adjoining the DPV's headquarters on the outskirts of Dakar, and its new laboratory was equipped, in line with the project's "ongoing policy," with additional, expensive laboratory equipment (including an HPLC for analysis and additional apparatus for sample preparation). The project had also received a gas chromatograph from a collaboration with the IAEA for its Special Program on Food Safety, which had selected the Locustox environmental chemistry lab as a site for the analysis of pesticide residues in produce imported and exported by Senegal. Indicators were also transported to the new site. The backswimmer *Anisops* was acclimatized to its lab conditions, and the report recommended the construction of artificial basins for fish testing organisms as part of a fourth transitional phase.[41]

The goal was for Locustox to evolve from foreign-funded, short-term (yet multiple and consecutive) public projects to a national institution with a durable national existence. How to do this also raised tricky questions about how ecotoxicological expertise, methods, and knowledge could be accumulated and deployed for public protection. Meerman had warned in 1993: "Environmental studies are not revenue-generating." In Europe and North America, these were regarded (presumably by governments) as "fundamentally important." Although no less important in Africa, it was simply not feasible there, Meerman asserted, without "external financing that is complimentary and durable."[42] This had been part of Meerman's plea for a third project phase, needed, he argued, to continue building up methods to attain regulatory validity. Yet his statement also cast doubt on the likelihood that either the states of the Sahel or private sources would be willing or able to maintain the availability, validity, and deployment of these evaluation and monitoring methods.

At the same time, Meerman had made suggestions for extending the legacy of Locustox beyond the end of the project. These included the institutionalization of ecotoxicology within the DPV as a "stable component" of its organizational chart, as well as the formalization of existing links with the CILSS training center for agricultural officers.[43] As for the sustainability of financing for ecotoxicology, he wrote: "We might imagine, for example, the creation of technical capacity and the infrastructure required to analyze residues in agricultural products destined for national and international markets."[44] An evaluation mission in 1996 took up this suggestion for the "auto-financing" of ecotoxicology. By contrast, however, it advised that the project "evolve towards a legally autonomous structure" rather than as a branch of the DPV.[45]

The question of durability in a context of limited public funding was marked by tensions between the long-term production of ecotoxicological knowledge as a public good and the marketability of tests that could generate revenue. It was hoped that public demands for environmental impact studies, in addition to a private market for the measurement of pesticide residues, would continue to tie commercial activities to the broader spatial and ecological exposure of Sahelian populations. A 1998 report to the Senegalese government recommended that the fourth transitional phase focus on improving semifield bioassays (tests of toxicity in biological indicators that seek to combine laboratory control with field conditions) to reduce the costs, while improving the validity, of environmental risk evaluations as "an important factor for the future semi-private center."[46] In the same year, a Senegalese private consulting firm evaluated both public and private markets for the planned center's services. It reported expressions of interest by bilateral and multilateral agencies for both environmental impact studies and regulatory testing.[47]

After its creation in 1999, CERES-Locustox continued to receive support from the Dutch government and FAO for a last three-year transition phase. By 2003, CERES-Locustox was said to have "the infrastructure, equipment and capability to provide scientific advice and information on ecotoxicology not only to the Government of Senegal but also to the other CILSS . . . member countries." The foundation had also carved out "a key position in the area of certification of agricultural exports to international markets."[48] In other words, while it had reoriented its activities toward revenue-generating services for the commercial agricultural sector, the new center maintained the legacy of Project Locustox as an available infrastructure for public science. Indeed, that year, Everts—who had returned to Locustox as a project manager in 1994 and overseen its conversion and semiprivatization—was awarded the FAO's "B. R. Sen Award," conferred yearly to "the field officer who had made the most outstanding contribution to the advancement of the country or countries to which he or she had been assigned," for "helping to establish a research and training centre in Senegal, in a field that was *virtually new* to the country."[49]

This delicate balance between the provision of commercial quality-control testing and a public mission of environmental protection seems to have been maintained over the following years. This was helped by the high levels of field activity (and presumably of foreign funding) prompted by new locust invasions in 2004. By 2010, however, the center was recovering from a period of financial and institutional crisis, and ecological fieldwork—associated with both basic research and public regulatory activity—had become a subject of nostalgia for biologists, who complained of being pushed to reconvert themselves into versatile experts in quality management. The environmental chemistry lab was filled with the activity of fulfilling contracts for agro-industrial firms, while office desks were covered with glossy brochures for an organization facilitating public-private partnerships to promote the competitiveness of export produce from Senegal. The center's survival strategy threatened, it seemed, to narrow the spaces and publics that its remaining capacity could reach. Yet there was still hope, especially with the arrival of a dynamic new director

FIGURE 4.4. CERES-Locustox, main building. Photo by Noémi Tousignant, Dakar, 2010.

general, that the legacy of a broader public environmental science could be reanimated. As his colleague put it: "Our mutation [from project to center] wasn't easy, but now . . . we are able to maintain our basal metabolism."[50]

INFRASTRUCTURING PUBLIC SCIENCE

Elsewhere, I called the insects enrolled as indicators for Sahelian ecotoxicology "infrastructure."[51] My point was to draw attention to cumulative ecological and methodological work as part of the creation of durable scientific infrastructure, and to point out how nonhuman organisms were involved in this process. Thus indicator species formed an axis along which the results of successive studies could be linked and condensed into tangible tools, such as amenable organisms, model ecosystems, rearing colonies, standard methods, and competent scientists. These tools produced regulatory information, allowing for accumulated data about pesticide environmental effects to be converted into hazard rankings for the CILSS and FAO. Thus indicators "remembered" the work performed over the time of the project (and even some done before it) to create a durable methodological and regulatory legacy.

Yet I also wanted to point out, as I do here, how closely this project of methodological infrastructure building was intertwined with the project of creating autonomous indigenous scientists and institutions for ecotoxicology in Senegal. In arguing for prolonged public funding for the Sahelization of ecotoxicology, Dutch scientists such as Everts and Meerman successfully mobilized an infrastructural argument about the temporal, epistemological, material, and sociopolitical conditions required to deploy ecotoxicology for environmental protection.

I thus enlist infrastructure as a concept for thinking through the relations between time, science, and the public good. Infrastructure is not always public, nor does it necessarily serve publics inclusively. There is a robust body of literature that documents the exclusions and violence of private or privatized infrastructure, describes how infrastructures are designed to serve narrow, exclusive constituencies,[52] and points to ways in which those living in the absence or margins of public infrastructural provision must "infrastructure themselves."[53] Yet there remains an ideal of infrastructure as redistributive, durable, and provisioning that scholars and activists can evoke to call for public investment in public goods. Only the public, as means and end, may be able to reach and sustain coverage over space and time, and across social difference, to make infrastructure a source of social solidarity, cohesion, and protection. Thus, for Bruce Robbins, the "smell of infrastructure is the smell of the public"; their decay is mutual, and so must be their renovation, as both a material and political project.[54] Ash Amin calls for a "politics of the staples" and of "skunkworks" as a politics of basic urban infrastructure that might suture together spaces of the city severed by inequality and remake it as a societal whole, but which also can "remember" unused capacity over time to protect urban inhabitants, collectively, against uncertainty and risk.[55] Without infrastructure, both the space and time of solidarity, welfare, and protection are at risk of falling apart. The "enchantments of infrastructure" also engage, as Penny Harvey and Hannah Knox have argued, the converse "possibilities of collapse, of dereliction, disintegration and abandonment."[56]

Those who, like Everts and Meerman, argued for the prolongation of Locustox not only bought time to make indicators. They also fought against the abandonment of zones of "relatively low economic significance" in the worldwide distribution of capacity to make knowledge and protect environments.[57] As Everts wrote in 1997: "In the hot arid zones of the world, ecotoxicological research is *in statu nascendi*."[58] These areas needed their own basic knowledge that could not be extrapolated from the bulk of available data or from "the methodological framework of ecotoxicology and its role in environmental policy making [that] often has been developed in, and applied only to industrialized countries."[59] In these mostly temperate countries, governments began, from the 1960s, to invest in data on the presence and accumulation of industrial toxins such as pesticides (initially the more persistent organochlorine compounds, then including shorter half-life molecules such as organochlorines and carbamates) and PCBs in ecosystems.[60] Prolonging Project Locustox can be seen as an attempt to catch up with such histories of public formation of ecological monitoring infrastructures.[61]

What was initially formulated as an epistemological argument for making the "Sahelian" indicators needed to evaluate the impact of locust control was gradually made into a broader call to to create the methodological and institutional infrastructure in the Sahel needed to alter the geography of ecotoxicology. As an infrastructural project, it entailed temporal aspirations to make cumulative and durable the knowledge gathered on local ecosystems and their alteration. Prolongation would crystallize in standard methods, experimental organisms, experienced scientists, well-equipped labs, and strong, networked institutions. In other words, the project stemmed from, and helped define, a broad vision of what scientific infrastructure entailed, and thus of what it would take to build it up. This vision was translated by the project's proponents into strong arguments for sustained investment in different infrastructural components, ranging from long-term ecological observation to the training of local scientists and the provision of equipment to national scientific institutions. The longer it lasted, and the more resources it absorbed over time, the more the project would be able to leave behind.

Although there was no explicit articulation of the connections between Sahelian ecotoxicology's methodological and institutional infrastructures, these were entangled both in justifications for prolonging the project and in the project's practices for accumulating data and making it usable over time. Thus, Project Locustox was successful in building capacity not only because it mobilized a broad understanding of infrastructure (and the resources necessary to create it) but also because of the ways in which its justification and practices integrated different infrastructural components as part of a single movement toward the institutional and epistemological future of ecotoxicology in the Sahel.

Yet the postproject survival of Locustox reveals the fragility of such integration and casts doubt on whether infrastructure, or capacity, can ever be durably "left behind" in the wake of the forms of support—financial, material, and ideological—that previously enabled its creation and maintenance.[62] Some infrastructural components may be malleable, such as the lab equipment that was repurposed to measure pesticide residues for commercial agriculture instead of for environmental assessments. Others are mobile, such as the techniques for using Sahelian indicators that could travel to other labs. But what seems particularly difficult to preserve, as interests and support shift, is the integration of institutional, material, and epistemological/methodological infrastructures as a cohesive unit. Achieving this integration in Senegal, or anywhere else, seems to require stable financial and ideological commitment to large-scale, long-term connections between the objects, subjects, agents, hardware, and institutions of science. In the case of Sahelian ecotoxicology, as Meerman hinted, only public, governmental interests in regulation, crop protection, and environmental protection at the level of populations and territories may be able to support such connections.

5 · Waiting/Not Waiting for Poison Control

In early 2010, the weekly staff meetings of Senegal's national poison control center (Centre Anti-Poison, or CAP) were held in a spacious but unfinished building. Mostly unpainted and partly filled with debris, the building sparked doubt about the future of poison control: would it grow into a vital "missing link" in the public health landscape, or would it fall into premature ruin?

"We have to be patient," sighed Alassane Diagne, the CAP's financial manager. He explained that construction began in 2006–2007, but that it was blocked by delays in payments to the public works agency contracted by the state to build and renovate health facilities. Diagne then launched into what seemed like a well-practiced complaint: "Me, I tell people, we managed to convince the authorities to have this center, but for the Ministry [of Health], it's a luxury, compared to malaria, AIDS, tuberculosis. I often say, the Ministry's cabinet director doesn't even know there are still snakebites [in Senegal]."[1] When we next met, however, he spoke another well-worn refrain: "We can't just sit there doing nothing" ("*Il ne faut pas rester les bras croisés*").[2] The CAP team had done a lot of the work "themselves," using flexible categories in their operating budget (which was also state funded but separate from payments for construction). By then,

FIGURE 5.1. Unfinished room, Centre Anti-Poison. Photo by Noémi Tousignant, Dakar, 2010.

a small wing of the building had been painted, electrified, and equipped with running water, office furniture, and a phone line.

The meetings took place in a narrow seminar room in this "self-finished" wing. Furnished with a long table, office chairs, air-conditioning, a Wi-Fi connection, and a few computers, the room doubled as a makeshift office for staff awaiting rooms in another part of the building. The meetings were presided by Amadou Diouf, the CAP's director as well as head of the toxicology unit at the university. He lost no time in pressing his staff to speed up the routinization of the center's activities. The priority was to get the phone helpline up and running. "It seems there isn't much these days," grumbled Diouf in early February.[3] They needed, he said, to start doing night shifts at home for now while waiting for a room. He urged the assembled staff to figure out how to get calls transferred to a cell phone. Who is volunteering to be on call, he asked? New phones had to be ordered, new lines if needed. At a later meeting, they discussed the scheduling system and the legislation needed to appoint and pay for extra interns, the design of the brochure to publicize the helpline number, and, when the finishing of the new rooms was imminent, night security and catering. Tentatively, they were working out an infrastructure of continuous presence.

The second priority project was a survey of snakebites—their frequency and consequences, how they were commonly treated, and with what available means. It was to be conducted, initially as a pilot, in a high-prevalence area in southeastern Senegal. Mathias Camara, the CAP's statistician, was in charge of the study. He explained to me that poison, as a cause of morbidity and mortality, was not (yet) a category in national health statistics.[4] Poisoning might be recorded in clinical registers, but not always; this was not an accurate reflection of the causes and magnitude of the problem. It was important, then, for the center to collect its own data. They would record reasons for helpline calls, of course, as well as compile incident notification forms as part of the CAP's "toxico-vigilance" (monitoring of poisoning) program. They also had to go further by initiating epidemiological surveys of poisoning. This was to begin now, without waiting for the state to finish building the CAP, equipping it with laboratory apparatus. Camara spent the following months painstakingly designing questionnaires for the snakebite study and planning how local health care staff would administer them.

The importance of generating data also arose in a lively discussion at one staff meeting about how much money to request for investigations to follow up on annual clusters of fatalities in southern Senegal thought to be caused by pesticide poisoning, but which the staff cautiously called "mysterious disease" (replaced, after discussing, with humor, more appropriate terminology, by "disease of unknown etiology"). A few members of the CAP had joined a team to investigate the deaths in 2005, but, though blood had been drawn, it had not been possible to confirm pesticide poisoning as the cause. Chastising the staff for underbudgeting the proposal, Diouf exclaimed:

> We need to collect data on the causal link between poisoning and pesticides. . . . We have to create an observatory, for long-term follow-up. . . . When you ask for money, you don't ask for the minimum. We already know there are problems with the use of pesticides. . . . We need to get blood samples, we need to get a spectro[photometer] and reagents. . . . We need field testing kits. . . . We need a sociologist too. . . . An epidemiological study for 2.5 million [CFA francs]? . . . We need 25 million! . . . We need data! . . . We will do everything. We will follow up, in a month, in a year. . . . We need right away to put in study *and* analyses. We have to follow up over a long period.[5]

After years of slow starts, since 2006 and even earlier, the CAP was, when I spent time there in 2010,[6] hastily setting in motion routine, long-term activity. "Not waiting" for the state, that is, not waiting for a building, lab equipment, or project grants, Diouf urged his staff to begin, as quickly as possible, to investigate and respond to poisoning on a continuous basis, for as long as possible, by answering the phone day and night, and by collecting data that would eventually represent toxic risk on the scale of the nation. This chapter situates this hasty routinization in relation, first, to the decades of slow starts that characterize the history of the project of poison control in Senegal since the 1970s. I then discuss how this project played out at a time of partial reactivation of public projects in the first decade of the millennium. This double historical positioning explores how toxicologists have ambiguously situated themselves in relation to the state. They have, on the one hand, attempted to be (or to become) the state as an orchestrator of a more expansive, more protective biopolitics of poison. On the other hand, they have also presented themselves as bypassing a slow, unpredictable, or ineffective state that seems incapable or unwilling to protect. In examining these changing ties between poison control and state capacity, I also reflect on the meaning of what, following Damien Droney, might be called the first (post-independence) and second (post-millennium) "ages of optimism" for public science and health in Senegal. Like Droney, who writes about a plant medicine research center in Ghana, I consider the ways in which this more recent age "is not a mere repetition of an earlier boom in hope for the future."[7]

SLOW STARTS

The CAP has had many beginnings: in the early 1970s, then around 2001–2002, and then again, in 2004, 2006, 2008, and 2010. Not everyone remembers, as Doudou Ba, formerly head of the university lab, does, that the idea of a poison control center has been around since "the 1970s or the 1980s." That's the time it takes, he explained, presumably referring to typical time frames for getting things done in Senegal or Africa more generally ("Generally it takes [at least] twenty years for a project!"[8]). The CAP's staff members are also keenly aware that their institution, like many other state-funded projects in Senegal, is vulnerable to uncertainties and delays. They also know that the current, still tenuous, materialization of poison control owes much to previous work by toxicologists, especially Diouf. Neither false starts nor moments of accumulating impetus, these many beginnings redirect a potential origin story of the center as born in crisis (of Ngagne Diaw) to a longer history of suspended plausibility.

In 1984, Michel Cailleux defended a thesis at the University of Paris (XI) that can be translated as "Rapid Identification of Solid Drug Forms (Tablets, Gel Capsules. . .) Authorized in Senegal: Contribution to the Creation of a File and of a Sample Library for the Future Anti-Poison Center of Dakar, in View of Treating Acute Drug Intoxications." In it, he asserts that Georges Gras first proposed a national poison control center in 1973, following a collective hydrogen sulfide poisoning (no further information is given or can be easily found about this event). It was, he writes, approved by the government of Senegal, but there was no follow-up. Yet the continued occurrence of poisoning regularly made, he asserted, the "need" for the center "felt." With press cuttings, he pointed to specific cases that had been reported in Senegal: of suicide by synthetic antimalarial drugs and one of the accidental consumption of food contaminated with parathion (an organophosphate pesticide). More generally, Cailleux pointed to the increasingly pervasive exposure of "*l'homme*" (humankind) to a "chemical world," of which the dangers were not yet widely known. Bemoaning the university toxicology laboratory's lack of usable equipment, he concluded with the hope that the poison control center of Dakar would be created quickly.[9] Until 2001, this is the last recorded or remembered mention of a poison control center in Senegal that I was able to find. For example, two theses on hospital poisoning statistics and pharmacovigilance, defended in 1984–1985, evoke neither the possibility nor the past proposal of a poison control center (though the second does mention a working group on a pharmacovigilance center that met in 1975).[10] Possibly, the suggestion was regularly brought up over the following decades but did not lead to anything concrete. Or perhaps, as the activities of existing public health institutions slowed or halted, no longer able to hire scientists or purchase equipment, the very idea became nearly impossible to take seriously and entertained, if at all, only as a joke or a fantasy.

In 2001, a working group began meeting at the national environmental agency (the Direction de l'environnement et des établissements classés, or DEEC) to develop an action plan for a poison control center. The plan would then be used to persuade the relevant ministries to support it.[11] The writing of an action plan was spurred by a program led by the United Nations Institute for Training and Research (UNITAR). It sought to build national-level capacity to meet the goals defined by the IOMC.[12] In the early 1970s, the proposal of a poison control center in Senegal could still be envisioned as part of a gradual, worldwide expansion of such institutions (which had begun only recently, in the 1950s). More broadly, it would have been coextensive with a multiplication of state regulatory agencies rising to the challenge of growing and globalizing chemical risks. By the early 2000s, however, poison control in the Global South, especially in Africa, had clearly become an issue of *capacity gap* (i.e., of temporal and spatial discontinuities in protective scientific and public health infrastructures that had not smoothly followed the spread of toxic risk).

Concerns with toxic protection as an issue of international coordination and capacity were, however, far from new. As early as 1980, the UN's environmental agency (UNEP), the International Labour Organization (ILO), and the WHO established the International Programme on Chemical Safety (IPCS).[13] By the mid-1980s, the global distribution and expansion of poison control or information centers was among its concerns. A 1984–85 survey did not list (on the basis of responses to a questionnaire sent out to a wide range of toxicological institutions) any poison control centers in sub-Saharan Africa.[14] The focus of a series of consultations and workshops into the early 1990s was on developing and diffusing policy and technical guidelines to help countries set up poison control services. That each country should have poison control mechanisms was also specified as part of the agenda for the "environmentally sound management of chemicals," adopted by the UN Conference on Environment and Development (UNCED) in 1992. Poison control—as an individual emergency service, but also as a system of detection of and response to collective exposure risks, and as a source of global data on poison (which would require harmonization)—thus became one of the goals of two post-UNCED bodies: the Intergovernmental Forum on Chemical Safety (IFCS), created in 1994, and the IOMC, created in 1995. While every country, faced with increasing toxic risks, needed poison control, specialized centers in the Global South would fill an especially large gap given the broader lack of data on poisoning and on the circulation of toxic chemicals. They could therefore play a particularly important role in the detection of collective exposures.[15] A series of international agreements on the control of toxic substances—the Basel, Rotterdam, and Stockholm Conventions, adopted, respectively, in 1989, 1998, and 2001—also increased the demand for capacity to detect and evaluate risk in the Global South.[16]

In conjunction with the IFCS and IOMC, UNITAR rolled out a series of programs in 1995, first to assist countries in assessing their existing infrastructure for "sound chemical management," and then, as a pilot in 1997–2000, to assist selected countries (Ghana, Indonesia, and Argentina, then Slovenia) in "Implementing National Action Programmes for Integrated Chemicals Management," which included, among other measures, the creation or consolidation of poison control centers.[17] In 2001, the program expanded to support Senegal in developing an action plan for a poison control center, as well as the creation of a Commission Nationale de Gestion des Produits Chimiques (National Commission to Manage Chemicals).[18]

By 2002, Diouf and Boubacar Camara, a pediatrician, had collected information for an action plan. The case for a poison control center, laid out in a slideshow presentation, rested on collated clinical data on poisonings (accounting for about 3 percent of admissions in a pediatric service, and 5.5 percent in a reanimation service), a list of existing threats (pharmaceuticals, pesticides, household products, etc.), and the inadequacy of existing response and surveillance capacity.[19] The center acquired its legal existence in 2004.[20] By 2008, when the Ministry of Health was alerted to a cluster of mysterious deaths in Ngagne Diaw, the center had a staff, an operating budget, offices in the ministry's main building, and a stalled construction site on the campus of Hôpital Fann, a teaching hospital adjacent to the university campus. Diouf and some of his staff had also visited poison control centers in Lille, France, and Rabat, Morocco. Meanwhile, at the university, Diouf arranged for an official agreement of scientific collaboration to be signed with the Université du Littoral-Côte d'Opale (in Dunkerque, France), home of the environmental toxicology lab he had been collaborating with for several years, and where analyses were performed for most of the studies copublished by the Dakar lab from 2003.[21] He also created a biotoxicology master's program, with which he could train some of the CAP staff while waiting for the center to become functional. In addition, this teaching program allowed the lab to obtain a percentage of student fees, with which they could invite collaborators from France and Morocco for lecturing visits. Sometimes, the French toxicologists carried frozen samples home (see chapter 1).

The CAP team's participation in the crisis response in Ngagne Diaw undoubtedly sped up this very slowly unfolding materialization of poison control. It was, the staff told me in 2010, an opportunity for the CAP to prove itself and to become more visible, by demonstrating the necessity and efficacy of poison control. First responding to the Ministry of Health's request to investigate the cause of fatalities, the CAP's findings helped bring in expert consultant teams from the Blacksmith Institute (which Diouf had previously worked with on air pollution and gold mining) and dispatched by the WHO (which included the head of the poison control center Diouf had visited in 2004). The CAP then represented the Ministry of Health as the national partner for parts of Blacksmith's remediation project (specifically, to monitor blood-lead levels and to conduct a public awareness campaign) and in a WHO-funded epidemiological investigation and treatment campaign (which involved screening for critically high blood-lead levels and managing the supply of therapeutic chelating agents). Crisis and external assistance activated a project that was not absent or dormant, but one, rather, that was expectant, waiting. This exceptional event dovetailed Diouf's longer-running struggle for capacity, data, and institutionalization.

In early 2010, the CAP was still starting. It now materialized as a half-finished, partly functional building, and as the first gestures of a set of routine activities. The center had no analytical equipment: the portable blood-lead testing kit from Blacksmith had broken down, and staff members were waiting for the NGO to provide a replacement to continue monitoring exposure in Ngagne Diaw. But they were clearly working toward a different sort of capacity, one that would allow for continuous responsiveness and data collection, and which would stretch well beyond the spatial limits of "toxic hotspots" and the acute time of emergency (see introduction). It would begin with the modest but continuous service of a telephone helpline and the first, paper-based steps (i.e., the filling out of questionnaires and forms) of nationwide data collection on toxic accidents—initially snakebite envenomation and adverse drug incidents (the CAP was also designated as a national pharmacovigilance center). Other projects were emerging, uncertainly, on the horizon: a long-term observatory of pesticides and toxic risks in Casamance, a study of accidental caustic soda poisoning in children (the prevalence of which was confirmed by the action plan investigation), a national program for the prevention and treatment of snakebites, and an international conference on envenomation by snake and scorpion bite, planned for the following year.

FIGURE 5.2. Centre Anti-Poison, unpainted, February 2010.
Photo by Noémi Tousignant, Dakar, 2010.

FIGURE 5.3. Centre Anti-Poison, painted, July 2010.
Photo by Noémi Tousignant, Dakar, 2010.

The university lab's members have described themselves as serving the state while complaining that the state does not give them the means to do so. The first is an instance of what Thomas Bierschenk describes as a "moral contractual relationship with the state," often cited by older bureaucrats. Younger ones, he notes, are likely to accuse the state of having contravened the terms of this contract.[22] As seen in previous chapters (see chapters 2 and 3), one of the lab's functions has been to provide medicolegal testing to the police and judiciary services. Ba remembered that lab members sometimes "had to use their own means" to perform these services. In 2010, there were no more means, and these requests were turned down. Ba and Babacar Niane also described the lab's research activities in the 1980s and 1990s as being of interest to "public authorities." During the same period, however, studies of food quality—of the conformity of butter and toothpaste, of potential contamination in peanut products and tea—were later remembered as work that government control laboratories "should" have done, suggesting the lab had stepped in for an ineffective state.

These contradictory statements are unsurprising. States everywhere are made up of heterogeneous entities, "servants," and institutions with various degrees of autonomy from, and identification with, the state, which, as a policy or an imagined ideal, is made discrepant by many factors, including financial limits, from its practices and effects. Yet where (national) state funding is particularly inadequate and erratic, as well as incoherent—paying salaries, for example, as has been the case for Senegalese toxicologists, but not for equipment needed for research and teaching—and where other resources (foreign, private, international) are routinely mobilized to enact governmental functions, but often only partially, it seems there can be particular confusion about *who* the state is and what it does, or does not do.

Such confusion is apparent in remarks made concerning the delays and accelerations of the concretization of poison control. By creating, staffing, and (partially) building the CAP, it seems that the Senegalese state wants to become a state that protects against toxic risks. Yet while this has generated some building and activity, it has also led to uncertainty and delay that threaten to make the center's partial existence a charade. Diagne recounted the CAP's initial response to the Ministry of Health's demand to investigate in Ngagne Diaw as a "gymnastics." "We had to act fast!," he said emphati-

cally. "There were dead people, there were sick people, they put pressure . . .
we had to act fast, show that we are efficient, reliable, competent, we had to
show we are resourceful [*débrouillards*]. The Ministry cannot wait, but mobilizing the state's means . . . it's very slow. . . . We had to be imaginative."[23]
Then Diagne went on: "Diouf often says, we have to invite the Minister [of
Health] to the center to show what we managed to do with so little money,"
explaining again how they "themselves" (but with a state-funded budget)
installed electricity, water, and office equipment. Diouf did indeed often insist that they must "not wait for the state" to get the CAP up and running.
Yet one day, Diouf also declared, with a sweeping gesture and, apparently,
pride, that "the state" had paid for all this, that they had not gotten anything
from "*bailleurs*" (donors). At the same time, I knew he was actively pursuing bi- and multilateral funding agencies. These contradictory statements
can be seen as akin to what Bierschenk called "doing the state, *en atten-
dant.*" What he describes is working (or pretending to work) for the state
while looking out for alternative employment opportunities. While this
meaning is applicable to some CAP staff members, I evoke it here to also
refer to working *as* the state while looking out for opportunities to obtain
additional funding beyond the state.[24]

According to Diagne, (people at) the ministry had explicitly advised
them to "get closer" to donors. And so they should, for Diagne said that
borrowing from the Ministry of Health was a "real pain" ("*la croix et la ban-
nière*"), so it was better to get other sources of support, like the WHO or the
World Bank. But at the end of that conversation, he said: "At the Ministry
of Health, they are not doctors. The politician is more successful than the
technician, is better at getting financed, has more contacts. Prof. [Diouf]
can identify [problems that need to be addressed], but he can't get the
means." These words may be vague, but it is no secret that external donors
fund a significant percentage of the Ministry of Health's budget. Diagne's
earlier remark, about poison control as a "luxury" for the ministry compared to malaria, AIDS, and tuberculosis, also alludes to the substantial
grants the state receives to address these diseases, in particular from the
Global Fund to Fight AIDS, Tuberculosis and Malaria (GFATM). This funding, then, is what defines the difference between necessity and luxury in
matters of public risk control and care. And so, perhaps, Diagne was asking: Why should the "politicians" not do their job of getting poison control
funded (by donors), and let the "technicians" do theirs?

Diouf and Diagne, who have probably done the most to get the CAP up and running, thus described the state in ambivalent terms, as an indispensable yet unreliable, and especially as an *incomplete*, partner to their efforts toward making the state continuously present as protector against toxic risk. They spoke of having to improvise with limited resources in order to convert slow, uncertain support into rapid, regular activity. This improvisation may be the first, hastened steps of an imminent expansion and routinization. Yet it also appeared, in some ways, merely as the performance of a nascent "bureaucracy of poison": some coats of paint and office equipment, the receipt of a few notification sheets of adverse drug events (exclusively from disease-control programs funded by the GFATM), some hours when calls to the emergency number would be answered (with much uncertainty about what to do next; the one call I witnessed, a potentially serious security risk that may have been a joke, was met with confusion). Uncertainty about future funding, from the state or potential donors, threatened to make this unfinished performance a barely convincing simulacrum of poison control.

Yet this performance also made the center exist, including as a target for future support. Diagne's comments about politicians and technicians notwithstanding, he and Diouf appeared as quite savvy in their efforts to market the CAP. They might be described not as "politicians" but rather, perhaps, as entrepreneurs of public service. They had worked together before, in the late 1990s, at the national drug control laboratory (Laboratoire National de Contrôle des Médicaments, or LNCM). There, they had adroitly handled a surge in state-channeled donor funding, setting up procedures of quality management to bring the lab up to global standards. At the CAP, they hired two journalists to run "information and communication" activities. As Diagne's aforementioned remarks suggest, they were also sensitive to the need to perform well for state authorities. As Diagne put it, "people have to feel that CAP exists and that it intervenes. The telephone helpline is the main activity, the core . . . that's what should animate the center. . . . When the telephone helpline works, everything else will follow." Meanwhile, Diouf and Diagne used their operating budget and personal connections to get a signpost made, adorned with a logo designed by a private public relations company, and brought in a consultant to discuss the possibility of implementing quality-management procedures at the CAP. Advice from a health consultancy firm was also solicited for a proposal to the Ministry of Health for a national plan on envenomation.

To what extent, and in what ways, did a half-built poison control center, the apparent resuscitation of a project that had died in the early 1980s, make manifest a new era of political and economic possibility in Senegal? Ambitious futures reentered the public arena in 2000, when presidential elections ended the forty-year reign of the Socialist Party. Senegal's first president, Léopold S. Senghor, stands for the rise and early fraying of confidence in the state's power to chart an accelerated course toward economic and cultural modernity (as well as the skillful maintenance of patronage ties). His handpicked successor, Abdou Diouf, commonly called a technocrat, guided the country, in the 1980s and 1990s, through economic crisis and austerity, and external pressures for economic and political liberalization. In 2000, Abdoulaye Wade, a long-standing prominent opposition party member, promised change (*Sopi*, the keyword of his electoral campaign) and, after the era of adjustment, a newly bold vision of how the state would make a more prosperous and powerful Senegal emerge.[25]

This vision took shape in what Ferdinand De Jong and Vincent Foucher have termed Wade's presidency's "appetite for infrastructure."[26] This included grandiose announcements that never materialized, such as a nuclear reactor, a free-trade zone, and a high-speed train network (some mocked these as the delusional fantasies of an octogenarian autocrat); others, such as the utopian "University of the African Future," were left half built.[27] Some projects, such as the refashioning of Dakar's highways and fashionable coastline, and especially the erection of a huge monument to "African renaissance," were highly contested and criticized, among other things, as frivolous spending of funds of dubious origins.[28] Yet even unrealized and controversial, the sheer scale of these projects appeared, De Jong and Foucher argue, after the curtailed horizons of structural adjustment, as an announcement of the state's return to the task of development and of future-making.[29]

This "return" indeed seemed to echo earlier thinking about the relationship between the state, infrastructure, and development, recycling a hopeful vocabulary of emergence, renaissance, and pan-Africanism. And yet, as Caroline Melly, as well as de Jong and Foucher, point out, these projects did not merely reiterate bygone aspirations but also sought to position Senegal as a global destination for foreign capital, donor funding, and tourism, which Wade actively courted (and which were joined by growing diaspora remittances and investments). Indeed, he has successfully kept Senegal one of the most "assisted" countries in Africa (but not the neediest). Not only were many projects built using a complex mix of state, international, and private means, but, by *performing* emergence, as "infrastructural spectacle,"[30] they sought to *conjure* emergence and consolidate the state's power to mobilize resources. De Jong and Foucher suggest that the Monument de la renaissance africaine be seen as a "postcolonial fetish,"[31] not simply marking but indeed summoning a massive transformation. They also point to the peculiar mixing of entrepreneurial spirit and (neo-)modernism with a tongue-in-cheek remark on Wade's claim to royalties on the statue: "The head of state as chief artist who defends his intellectual property rights, strange Stalinian echo in a neoliberal era!"[32]

Thus the half-finished cap can be seen, on a very general level, as dovetailing a bigger boom in construction that was part assertion of the state's renewed centrality in orchestrating "emergence," part speculative investment (for the state itself and for the private capital it sought to attract).[33] More specifically, however, this project also took shape during a time of increased spending on health in Senegal. From 1998, when Senegal, under Abdou Diouf, established a strategic plan for health development (the Plan national de développement sanitaire), to 2007, the health budget grew from 37 billion cfa francs (about 63 million usd at the 2016 exchange rate) to 140 billion cfa francs (about 2.4 billion usd at the same rate). Just over 50 percent, on average, of this budget has been paid for by the state, with significant increases from 2000 to 2002, and again around 2006. Local government and "populations" also pay for health services: the latter's contributions more than doubled, from 7.8 to 17.8 billion francs, between 2000 and 2002, and rose to 28 billion in 2004, thus showing a clear privatization trend. "External partners" (i.e., bi- and multilateral donors) accounted, on average, for 27 percent of the budget, increasing markedly in the late 1990s and in the period of 2004–2006.[34]

The contributions of "external partners" have been particularly visible in national disease-control programs focusing on the "Big Three" of global health: hiv/aids, malaria, and tuberculosis. This new influx of resources is tied to a worldwide exponential rise in global health institutions and resource flows, especially for these diseases, as well as vaccination and maternal and child health interventions. A growing body of work in anthropology explores the complex effects—on responsibility, provision, and citizenship—of global health resources, imaginations, and modes of accountability. Some, looking at health programs that operate outside national infrastructures, or at the imperatives that guide the planning and provision of global health technologies, have described an emerging "biopolitics without the state," where citizenship—or rather humanity, insists Tobias Rees—and government are reformulated in terms of "life in crisis."[35]

Others, however, have instead traced a remaking of the state as mediator of global health, as it manages its multiple relations of accountability to "stakeholders," developing strategies to keep resources flowing while projecting its image as a provider of care and opportunity. Indeed, studies from Senegal, where the state has kept tight control over donor-funded programs and increased its own health spending (as well as the contributions made by its people), show that national and global imaginations of citizenship have become tightly imbricated.[36] This does not mean that the public of public health is defined and cared for inclusively as a national citizenry. The allocation of global health resources to priority problems (those calculated as posing the greatest "global burden") has selectively reanimated nodes of national clinical and laboratory infrastructures while motivating the creation of new hybrid "para-statal" (trans)national and public-private institutional forms.[37] Along with the growth of private clinics, the landscape of health research and care has thus been fragmented into spaces of "abundance and scarcity,"[38] creating "archipelagic" geographies of globally connected and well-resourced "islands" that are disengaged from contiguous populations and territory.[39] At the same time, "islands" such as malaria or HIV treatment programs can have broader, unanticipated impacts in terms of how citizens (patients, health care workers, hospital managers, private pharmacists) make claims on the state, and, conversely, respond to the demands they are subjected to.[40]

During this time (from the late 1990s, but largely under Wade's presidency from 2000) of pursuit of global health funding, private investment, public-private partnerships, and, more abstractly, of an ethical/aesthetic blend of public care and entrepreneurship, toxicological institutions have been reanimated in various, but very partial, ways. While the university laboratory has had hardly any functional analytical apparatus since the early 2000s, it has developed new sources of revenue and research activity. The lab now offers two short diploma-granting courses, publicized in color brochures sporting the lab's new sleek logo. These produce both marketable skills and money to spend on consultations fees and office equipment. This is made possible by university-wide market-based reforms (implemented in many African universities) that respond to a trend of "massification"—rapidly rising student numbers (at UCAD, student enrollment grew by 25 percent between 2001 and 2006) not matched by corresponding increases in staff, space, or equipment—by creating ways of bypassing constraints on tuitions fees for regular degree programs. Meant to attract students from professional, business, or NGO environments, these programs increase the private financing of a public university, with units/departments obtaining a percentage of fees, thus giving them an incentive to make these "attractive."[41] Meanwhile, the lab's partnership with the Dunkerque laboratory, and the flow of newly hired junior staff (as PhD students, funded by scholarships from the Agence Universitaire de la Francophonie), visiting scientists (funded by biannual state-funded travel bursaries), and consultant lecturers (funded by the diploma programs), has led to a rise in the number and quality (more precise and complex indicators of exposure, higher-impact journals) of its publications.

Recall, also, that Project Locustox was semiprivatized in 1999, and that although it has received public funding, especially from the FAO, to conduct environmental evaluations and participate in pesticide-reduction programs, its bread-and-butter activity has, in recent years, been the analysis of residues on produce destined for export markets. Wade's government has promoted various initiatives to increase the volume and competitiveness of agricultural exports, notably by stimulating public-private partnerships, through the Programme de Développement des Marchés Agricoles du Sénégal (Program for the Expansion of Senegal's Agricultural Markets, or PDMAS), funded by various donors, notably the World Bank, and launched in 2004. According to the World Bank's website, as of April 18, 2017, the volume of exported produce had, by 2010, grown from 13,000 to 30,000 tons. Quality management and certification is one of the subcomponents of this program.

Although the LNCM is not a toxicological institution per se, it has been run by members of the toxicology lab (Amadou Diouf, then Mounirou Ciss) and addresses a problem—that of poor-quality (fake, substandard) drugs—that has often been defined in terms of its poisonous effects (i.e., uncontrolled medicines kill). The LNCM's history of multiple beginnings, suspension, and reanimation in some ways parallels that of the CAP's: although it gained legislative existence earlier (in 1979), it largely remained in a state of suspension in the 1980s and 1990s, obtaining some staff and equipment in 1987 but remaining at a low level of activity.[42] From 1998, however, the lab was moved to a prime location in central Dakar and re-equipped. Ciss insisted, when he gave me a tour of the facilities, that the state had paid for *all* of its extensive, cutting-edge equipment. And yet, as several staff members told me, this apparatus was activated almost only to perform tests requested by the HIV/AIDS, tuberculosis, and malaria programs. By contrast, the pharmacy division of the Ministry of Health did not put in requests or provide resources to regularly test batches of other types of medicines coming into the country. Although the LNCM has apparatus, it is difficult for it to obtain reagents and reference substances other than those provided by these heavily externally funded disease-control programs. Indeed, grants from sources such as the GFATM come with both requirements and earmarked funds for drug-quality testing that can pay for lab consumables—this is an example of the ramifications, for national public systems, of the forms of regulation and monitoring built into global health funding.[43] The LCNM has received support from other external sources such as the WHO, the European Union, and USAID, which is at least partly tied to strengthening capacity to certify batches of yellow fever vaccine produced in Senegal—at the Institut Pasteur, just up the road from the new LNCM—in order to release them for export, but also, again, to the monitoring of drugs used in the national disease programs. In other words, what appears as a well-equipped and functional state institution is not, in fact, provided with the authority and resources to monitor the safety of all types of drugs that circulate in the country, and thus to protect the public at large from the risks of substandard medicines.

Thus the CAP appears as one of several partly built or partially reactivated public health institutions, which, if we take Diagne at his word, the state has funded in the hopes it will attract additional private or donor resources. Diouf's insistence on "not waiting" appears as an effort to hasten the extension of poison control, to *be* the state even before the state manages to deliver a finished building, lab equipment, and research funding. Yet it can also be seen as an advertisement, that is, as a performance of the CAP's potential to protect, enhanced by an attractive logo and signboard, and perhaps the implementation of quality-management procedures. There is no reason to doubt either Diouf's commitment to public service

FIGURE 5.4. Visitor entrance, dust-covered and not yet in use, Centre Anti-Poison. Photo by Noémi Tousignant, Dakar, 2010.

or his good sense, that is, his realistic assessment of needs and opportunities for funding. Yet this ambiguity underscores the fragility of the CAP's partial existence in 2010. Without the state, its future survival was insecure given that it was not seeking private clients, and poisoning was not yet recognized as an obstacle to economic growth or a global health emergency.[44]

Epilogue · Partial Privileges

There were times when I wished I were writing another book. Models beckoned. There were compelling histories, set in the United States, in which toxicology clashed with or eluded other ways of defining the presence and meaning of toxicity, such as Linda Nash's *Inescapable Ecologies*, Michelle Murphy's *Sick Building*, or Chris Sellers's *Hazards of the Job*. There were also hard-hitting histories, such as Jerry Markowitz and David Rosner's *Deceit and Denial*, in which the obstruction of protective knowledge by industrial interests was described in minute and lurid detail. Set in Botswana and Senegal, there were moving ethnographies in which the improvisation of (self)-care in spaces of neglect was set alongside vivid accounts of illness and well-being, such as Julie Livingston's *Improvising Medicine* and Duana Fullwiley's *The Enculturated Gene*. Set in India and Ukraine, there were harrowing ethnographies, such as Kim Fortun's *Advocacy after Bhopal* and Adriana Petryna's *Life Exposed*, of the aftermath of toxic disaster, in which the meaning of exposed bodies was negotiated alongside the locus of protective responsibility and claims to care in liberalizing, postcolonial political economies. Set in India, David Arnold's spellbinding *Toxic Histories* takes us along the thread of poisons into the fabric of the everyday, the panics, evidence, horrors, nuisance, and hopes of a colonial society.

From the world of toxicologists, I caught tiny but tantalizing glimpses into how "the poisoned poor" themselves defined the corporeal and economic stakes of handling pesticides, batteries, and waste, and of why contamination rarely penetrated, and more often was absent from, national and global agendas. Intrigued by these glimpses, conference and seminar audiences wanted to know more: about "indigenous" ontologies and epistemologies of poison, about the anxiety and suffering elicited, or not, by substances and situations that toxicologists defined as risky, about why toxic exposure was not (yet) seen as an issue of global health and national government (the implicit question, which was not asked, was about the state's prioritization of economic growth and industrialization). In other words, they asked: What do the "exposed" and the "at risk," on one side, and, on the other, the purse-string holders (and possible accomplices of evil capital) think about poisons? Were these not the real subjects and actors of (un) protection? Couldn't I somehow supplement my story about toxicologists with more detail from these other vantage points?

Part of the answer as to why I did not say more about these other views on poisons is methodological. These other perspectives on the stakes of exposure and contamination have only occasionally and briefly intersected with the work of toxicologists or shaped, in significant ways, its meaning and possibilities. In the meantime, the concerns of "the people" and "the state"—about work, dangerous substances, misfortune, the export economy, pollution, and so on—have not generally been identifiable as specific concerns about the presence, level, and pathways of toxic elements and molecules in Senegalese bodies and environments. In other words, contamination and exposure have been most visible in, and indeed *made* visible by, toxicologists' work. The "at risk/exposed" have come into view as such mainly when toxicologists have been involved, in response to mysterious deaths or in calls for further monitoring following modest studies pointing to actual or potential problems of contamination. Exposure has not, in any immediately identifiable way, come up as a topic of public protest, debate, or activism (though risky matter, such as e-waste and pesticides, have been the target of modest activity from the Dakar-based international NGO Enda Tiers Monde and the local chapter of the Pesticides Action Network, for example). And, as I have shown throughout this book, specific state and international investments in toxicological capacity have, for the most part, been modest, brief, and far between. I have, to some extent, pulled on the thread of these investments to sketch out a broader picture of international interests in toxic risks to health and the environment, and of how the Senegalese state bets on problems of global health and the export economy. I do not, however, claim to have fully elucidated the logic by which toxicological research and regulation has been only partially capacitated. My main concern instead has been to describe what has happened to toxicology and toxicologists between, after, and beyond these investments and involvements, and to describe this work as the most continuous and, in some ways, the most expansive and ambitious engagement with issues of toxic risk in Senegal.

Thus the histories of toxicology, exposure, and funding choices could not be told together through their points of contact. Moreover, it was also difficult to research them in parallel. To describe toxicologists' work over time has been a laborious undertaking, for there was no predefined archive or body of documents. To get sources, I had to form relationships with toxicologists and their institutions, through which I might create, locate, access, and spend time with their memories, piles of unpublished material hidden away in drawers and cabinets, and equipment in various states of repair. This was not compatible with what would have been an equally laborious, necessarily ethnographic, and worthwhile endeavor: to study how a community identified as exposed or at risk interpreted the presence and effects of toxicity—and the stakes of its diagnosis—amid their other concerns with survival, well-being, suffering, and misfortune (Ngagne Diaw being one of the few obvious candidates for such a study). Even government records are not easily accessible; much is not archived or is still kept—if at all—by ministries or individuals, to which access must be negotiated. More detailed histories of, say, the IOMC or of the Blacksmith Institute may have been less time-consuming but still difficult to reconcile with—and not directly relevant to—a long-term and close-up view of the aspirations and capacities with which they intersected in the first decade of the 2000s.

The question remains as to why toxicologists are at the center of this study. I came to them not, initially, from an interest in toxic hazards but from an interest in pharmacists. Earlier, while researching pharmaceuticals, I became both frustrated with lacking and uninformative archival sources, and at the same time particularly interested in what happened to colonial problems of therapeutic inaccessibility, economic/regulatory dependence, and scientific expertise in the post-independence decades. Pharmacists, who, before, but especially after, independence were involved in drug policy and supply, expanding distribution networks, medicinal plant research, quality control, initiating local pharmaceutical production, and so on promised richer sources of information about this period.

They could also refocus my research on people, action, and meaning rather than on things. It was by chance that two of the first pharmacists I met, through contacts, during a preliminary research trip were Doudou Ba and Mounirou Ciss (also two other important and related studies were under way: Kristen Peterson's work on drug markets in Nigeria and Donna Patterson's on private pharmacists in Senegal).[1] I decided to pursue this lead mainly because there did not seem to be any history or ethnography of toxicology in Africa nor much scientific or scholarly engagement with the issue of contamination, and also because of a rather unfocused fascination with the themes of risk and quality that this path of research seemed to open onto.

It was only gradually, in the course of my research and writing, that I got a clearer idea of the story there was to tell and of why it was important. My answer is woven into the details and interpretation that fill the pages of this book. Toxicologists' past tells us that contaminants did not suddenly appear as a problem or concern when children began dying in Ngagne Diaw in 2007, or when UNITAR/IOMC suggested that an action plan for a poison control center be drawn up in 2001. Many other scientists in Africa have surely been quietly, modestly working to define problems of public concern in studies that appear only in local medical journals or student theses, or that never get started or finished at all. We need histories not just of scientific problems and knowledge that cross paths with African scientists but also fine-grained, uninterrupted views of what it has meant to be a scientist in Africa, both on and off the trajectory of international "capacity-building" and of global histories of science and medicine. These stories can tell us what capacity is, how it is made, kept, and lost, in relation to the specific stakes of what existing or missing capacity might be used to act on, in other words, in relation to what "good science" in both epistemological and moral terms might be.[2]

Yet my commitment to this story also, in part, grew out of a more personal reflection about what it was that made me uneasy and confused while living and working in Dakar among people who "should have been" but were somehow not quite, just like me. My neighbors and extended family did not go hungry, could afford medical care, had decent housing, and dressed well. They were, apparently, middle class. Yet lack of access to credit, insurance, and enough stable or well-paid jobs, combined with gaps in public services and benefits, and the webs that tied them to even more precarious circles of relatives and friends, seemed to shrink horizons of expectation, lower the threshold of luxury, and allow destinies to take sharp, sickening turns. The neighborhood on weekdays was filled with listless, qualified but unemployed young men and women of all ages, sometimes joined by children whose school fees could not be paid for the time being, or neighbors and relatives in search of a free midday meal, or cars parked for lack of gas money. Cuts in the water or power supply were unpredictable. The imported, processed foods that were eaten routinely (cookies, pasta, powdered milk, mustard) seemed incredibly cheap and of dubious quality, while fruit, UHT, milk or plastic pouches of yogurt were occasional treats. Most people had less furniture and flimsier stuff than I was used to, and no books. Bills and coins were constantly being hidden in handshakes, to reimburse a friend or relative for "transport," or to help out with this or that. Some recipients were just as well dressed, just as well educated, perhaps recently just as well employed as the giver.

Like wealth and comfort, life seemed unexpectedly fragile. The death of two nieces in my family-in-law shook me. One was admitted to hospital in a coma with what looked like but was undiagnosable as meningitis and died a few months later. Another, who my daughter played with often, died only two or three days after being diagnosed with leukemia. An uncle and a childhood friend of my husband's both died, apparently from cancers, although they had been diagnosed only months earlier with tuberculosis and diabetes; another childhood friend disappeared even more mysteriously. The long-ago death in childbirth of my mother-in-law—who, again, could afford transportation and care—haunted both my pregnancies. I do not know enough about any of these deaths to say just how preventable they might have been. But they felt wrong: too quick, too many, too unexplained. And these were all people who had the contacts and money, of their own or from close relatives, to obtain what treatment and care was deemed necessary.

The people I spent time with in my field sites were even more familiar. They had salaries, cars, nice suits; it was not uncomfortable to be invited to lunch at the Faculty Club. And we were fellow researchers, as they pointed out, to emphasize their commitment to giving me time and access. Yet here I was, doing what I was trained to do, but they, it seemed, felt they were not. They kept busy, were successful, and got promotions. But they were waiting . . . for the next project, the next trip overseas.

They wished they had a working lab, were producing knowledge more regularly, could go out and "get a plant and determine what is in it," as Niane had put it, publish more in higher-rated journals, have their students do practical exercises and lab-based thesis research, or at least run or inform functional public health institutions. They wanted to be better scientists for the pleasure and status of it, of course, but it also seems obvious that if they could do their jobs "properly," they would also be taking a step toward identifying threats to public safety and well-being.

The absence of the state, of science, and of regulation as a protective presence in Africans' lives does not happen only "out there" in the global political economy, or "up there," when a higher official concentrates scarce resources on something—expected to drive up indicators of economic growth and mortality decline, to please the Global Fund or the World Bank, to obtain political loyalty or kickbacks—and neglects something else. It also happens in a space of partial privilege, in which public servants and experts who consider that it is their job to define and control risks to the public are not (quite) able to do so, and in which middle-class people can easily lose the ability to pay for their own individual protections and are not covered by a finer-meshed net of collective protections that some of us who live in only somewhat dismantled welfare states still take for granted. It seems that both academic and popular audiences outside Africa know little about these positions of partial privilege. These are not the most unprotected positions, in which so many Africans who struggle to meet basic needs (and which are unquestionably worthy of the greatest share of scholarly attention). And they may well perpetuate the inequalities they incompletely benefit from. As Thomas Bierschenk remarks, early work on public services in Africa focused on corruption, assimilating workers of the state to its broader process of privatization and criminalization.[3] Yet his team's subsequent project found "another face of the creeping privatization of public services": when local officials use their own private resources to make calls, get to meetings, and transport and feed prisoners, in other words, to "make the state work."[4]

This book describes acts analogous to these, acts not of activism or of heroism but merely of efforts to do one's job, and yet which bring a minimum of public service and protection into being. At the same time, toxicologists' bigger but seemingly banal wishes—to have their own lab(s), to inform regulation that works, to run a functional poison control service—

should also make us ask ourselves, whatever the condition of the welfare system under which we live and the degree of privilege we have within it: How much can we and should we expect from the state and from science? What protections and what possibilities for doing meaningful public work should be demanded, for and by Africans who, on the whole, have fewer of these, but that many of us, African or not, are at risk of losing?

INTRODUCTION

1   WHO, "Intoxication au plomb à Thiaroye sur mer, Sénégal"; Wilson, "Technical Expert Mission"; Blacksmith Institute, "Project Completion Report"; Haefliger et al., "Mass Lead Intoxication from Informal Used Lead-Acid Battery Recycling in Dakar, Senegal."

2   The competition is held annually by the Caen Memorial Centre for History and Peace. Fall's plea, "Ngagne Diaw ou le dernier mohican de Thiaroye," which won first prize in 2010, can be found on the institution's website: http://www.memorial-caen.fr (accessed August 11, 2015). Winning pleas in other years have addressed topics such as the death penalty, euthanasia, and women's rights.

3   Mourre, *Thiaroye 1944*.

4   Means, "Toxic Sovereignty."

5   The Blacksmith Institute is presented on its website, which was still live in October 2016. In 2015, it changed its name to Pure Earth but continues to define itself as "an international non-profit organization dedicated to solving pollution problems in low- and middle-income countries, where human health is at risk"; see http://www.pureearth.org/what-we-do/, accessed August 12, 2015.

6   Blacksmith received support from the Green Cross Switzerland and Terragraphics Environmental Engineering for this intervention.

7   See, for example, Blacksmith Institute and Global Alliance on Health and Pollution, "The Poisoned Poor."

8   Blacksmith Institute, "Project Completion Report."

9   The actors I followed in this study are those who have worked in or with three toxicological institutions and include individuals who were not trained as toxicologists (e.g., statisticians, managers, and entomologists, as well as the analytical chemists who work alongside toxicologists at the university) as well as scientists who are not Senegalese nationals (French technical assistants in the 1970s to early 1980s and Dutch ecotoxicological scientists who worked at Project Locustox in the 1990s). Yet all have worked toward making, keeping, and mobilizing scientific and regulatory capacity to monitor toxic risk in Senegal. The general term *toxicologists*, then, covers this group as a whole.

10 For a description and examples of Africanist, historical, and ethnographic approaches to scientific and health capacity, see Geissler and Tousignant, eds., *Special Issue: Capacity as History and Horizon*.

11 See, for example, Fortun and Fortun, "Scientific Imaginaries and Ethical Plateaus," and Boudia and Jas, eds., *Toxicants, Health and Regulation*.

12 In *Speculative Markets*, Kristen Peterson argues, somewhat similarly, that the regulation of fake drugs in Nigeria is largely futile despite significant institutional and technical regulatory capacity due to the structure of both national and international pharmaceutical markets, as well as the adjustment of African economies and the multiple economic and political pressures on local regulatory action. Her focus, however, is on the speculative practices that have created a vigorous and ultimately unregulatable pharmaceutical market, while I focus more narrowly on the specific limits placed on potentially protective toxicologically testing and research.

13 The term is from Mitman, Murphy, and Sellers, *Landscapes of Exposure*.

14 Boudia and Jas, eds., *Powerless Science?*

15 Roberts and Langston, eds., "Toxic Bodies/Toxic Environments"; see especially Daemmrich on biomonitoring, 684–693.

16 Beck, in *Risk Society*, characterizes late modernity as generative of pervasive, incalculable risk (including the production and release of hazardous chemicals) and as cognizant of the limits of scientific control. Boudia and Jas, *Toxicants, Health and Regulation*, 2, describe a late twentieth-century shift in the politics of contaminant regulation from the objective of mastering to that of coping with generalized contamination, a condition that results from the failure of regulation. They follow up on the failure of regulation as the result of its contradictory logics (to legitimate production and protect populations) in *Powerless Science?* The best overview of how different theories of modernization address the issues of unequal exposure to and protection from toxic risk is Pellow, *Resisting Global Toxics*, 18–24.

17 Pellow, *Resisting Global Toxics*; see also, on the relationship between industrial hazard and globalization, Sellers and Melling, eds., *Dangerous Trade*, and Sellers, "Cross-Nationalizing the History of Industrial Hazard."

18 Pellow, in *Resisting Global Toxics*, insists on the potential of activism to temper the maldistribution of risk and to reinforce protection (against theories of modernization that posit risk as inevitable) but pays little attention to the scientific study of sources, patterns, and consequences of contamination and exposure.

19 See, for example, Markowitz and Rosner, *Deceit and Denial*; Sellers, *Hazards of the Job*; Murphy, *Sick Building Syndrome*; the essays in Roberts and Langston, eds., "Toxic Bodies/Toxic Environments"; Frickel, *Chemical Consequences*; and Fortun and Fortun, "Scientific Imaginaries and Ethical Plateaus."

20 Hountondji, "Scientific Dependence in Africa Today," 9.

21 Brooke, "Waste Dumpers Turning to West Africa."

22 Robert Bullard's work on the contamination of black communities in the southern United States popularized this term; see his now-classic *Dumping in Dixie*.

23 Marbury, "Hazardous Waste Exportation."

24 Fortun, *Advocacy after Bhopal*.

25 O'Keefe, "Toxic Terrorism." See also Third World Network's collection of essays, *Toxic Terror*.

26 Cited in O'Keefe, "Toxic Terrorism," 87. Brooke, "Waste Dumpers Turning to West Africa," reports that an editorial in *West Africa* described the content of "dozens of letters from angry readers" and summarized the "traumas cited" as slavery, colonialism, and unpayable foreign debt.

27 A copy of the leaked memo, "Whistle Blower's Corner/Lawrence Summer's 1991 World Bank Memo," is available on the website of the Basel Action Network at http://ban.org/whistle/summers.html, accessed August 12, 2015.

28 Nixon, *Slow Violence*, 2.

29 Ferguson, *Global Shadows*, 71–78.

30 Livingston, *Improvising Medicine*, 30–31.

31 Clapp, *Toxic Exports*, 2–3, asks why, despite the outrage the memo provoked, international regulation has failed to control the circulation of hazardous waste and points to loopholes in this and later conventions—which she describes as a "leaky dike." One such loophole concerns waste labeled "for recycling" rather than disposal. While an amendment to the Basel Convention was proposed in 1995 to explicitly ban the export of waste for recycling from OECD to non-OECD countries, it had not yet been ratified or implemented by many countries.

32 The Rotterdam Convention concerns prior informed consent (specifying the criteria for a country to make an informed decision about allowing the entry of "certain hazardous chemicals and pesticides"), while the Stockholm Convention aims to restrict or eliminate the production and use of certain substances classified as persistent organic pollutants (POPs), such as DDT and PCBS, which remain in environments and bodies for very long periods of time. See UNEP, "The Hazardous Chemicals and Wastes Conventions," September 2003, http://www.pops.int/documents/background/hcwc.pdf.

33 Amnesty International and Greenpeace Netherlands, *The Toxic Truth*, 2.

34 Amnesty International and Greenpeace Netherlands, *The Toxic Truth*, 2.

35 Koné, "Pollution in Africa."

36 Stoler, "Imperial Debris," 197 and 204.

37 Means, "Toxic Sovereignty."

38 On African expectations of "convergence," that is, of full global economic and political membership in the future of development, see Ferguson, "Decomposing Modernity."

39 Lincoln, "Expensive Shit," 3–4.

40 Tadjo, "Dessine-moi (écris-moi) une independence . . . ," 66–67. Unless otherwise noted, all translations are my own.

41 On people, likewise, as waste of the economy, see Bauman, *Wasted Lives*.

42 Nixon, *Slow Violence*, 4.

43 Nixon, *Slow Violence*, 3 and 15.

44 Nixon, *Slow Violence*, 16.

45 Livingston, *Improvising Medicine*, especially 29–51; Hecht, *Being Nuclear*, especially 183–212 and 259–286.

46 It should be noted, however, that the Blacksmith Institute's activities in Africa are not limited to emergency remediation operations but also include education initiatives to make risky activities safer or provide individuals with alternative revenue-generating possibilities (e.g., this was the case in Ngagne Diaw) and, more recently, a toxic sites identification program to screen potentially contaminated areas for health exposure risks (thereby beginning to expand the map of known or partly known risks). Yet it remains focused on well-delimited "toxic hotspots" and emphasizes the need for urgent action in "the world's most polluted places." See the Pure Earth website and Pure Earth and the Global Alliance on Health and Pollution "The Poisoned Poor."

47 Overviews of sources of heavy metal pollution in Africa are given by Nriagu, "Toxic Metal Pollution in Africa," and by Yabe, Ishizuka, and Umemura, "Current Levels of Heavy Metal Pollution in Africa." A comparison of these reviews, published in 1992 and 2010, shows an increase in concern with, and studies of, metal contamination on the continent. The earlier report also shows that, despite the scarcity of data available at that time, this was already identified as a serious and worsening problem, especially for children in urban settings, although it should also be noted that this was prior to the phasing out of leaded gasoline.

48 See, for example, Diouf et al., "Environmental Lead Exposure: A Pilot Study"; Diouf et al., "Environmental Lead Exposure: A Cross-Sectional Study"; and Tuakuila et al., "Blood Lead Levels in Children."

49 See, for example, Chindah et al., "Distribution of Hydrocarbons and Heavy Metals," and Emoyan et al., "Evaluation of Heavy Metals."

50 See, for example, UNEP and Kimani, "Environmental Pollution and Impacts on Public Health." Cabral et al., "Low-level Environmental Exposure," is one of the rare studies in an African setting that tests for markers of exposure and its effects on the body rather than merely measuring levels of contamination. Asante et al., "Multi-trace Element Levels," presents itself as the first study of human exposure to e-waste recycling. See also Cabral et al., "Impact du recyclage des batteries."

51 See, for example, Lo et al., "Childhood Lead Poisoning"; Ikingura and Akagi, "Monitoring of Fish and Human Exposure"; and Van Straaten, "Mercury Contamination."

52 See, for example, Banza et al., "High Human Exposure," and Ikenaka et al., "Heavy Metal Contamination."

53 See, for example, Sobukola et al., "Heavy Metal Levels"; Gras and Mondain,

"Problème posé par l'utilisation des cosmétiques mercuriels"; and Obi et al., "Heavy Metal Hazards."

54   Kinyamu et al., "Levels of Organochlorine Pesticides Residues"; Diouf et al., "Utilisation des feuilles de manguier comme bioindicateur."

55   See, for example, Diouf et al., "Étude du niveau de pollution de l'eau de puit"; Pazou et al., Organochlorine and Organophosphorous Pesticide Residues"; Bempah, Kofi, and Donkor, "Pesticide Residues in Fruits"; and Anderson et al., "Passive Sampling Devices." While a few thesis research projects in Senegal (in both the toxicology lab and the plant biology department of the Faculty of Sciences) had, earlier, tested vegetable, fish, and water samples for pesticide residues, the first larger-scale, published study of pesticide levels in vegetable crops, confirming impressions that contamination levels exceeded accepted safety standards, only came out in 2016; see Diop et al., "Monitoring Survey of the Use Patterns and Pesticide Residues."

56   See, for example, Touré et al., "Investigation of Death Cases."

57   Williams et al., "Human Aflatoxicosis in Developing Countries."

58   Tagwireyi, Ball, and Nhachi, "Poisoning in Zimbabwe."

59   Amadou Diouf, oral presentation on pesticides, World Social Forum, Dakar, February 2011.

60   Banza et al., "High Human Exposure."

61   Anderson et al., "Passive Sampling Devices."

62   See "Completed Projects," Pure Earth website, http://www.pureearth.org /projects/completed-projects/, accessed October 17, 2017.

63   See "Global Monitoring Plan," Stockholm Convention website, http://chm .pops.int, accessed October 17, 2017.

64   Even a quick review of recent publications on toxic contamination and exposure in Africa shows that most rely on collaboration with foreign labs, where much of the testing is probably done (which means freezing and shipping small batches of samples); many are "pilot" or "exploratory" and comment on the scarcity of available data (especially on human markers of exposure).

65   For a good overview, see Gaillard, Hassan, and Waast, "Africa." The best overview for health is Prince, "Situating Health and the Public."

66   For example, medicinal plant science was funded as a matter of national pride and self-sufficiency in Ghana in the postindependence decades; see Droney, "Ironies of Laboratory Work," and Osseo-Asare, *Bitter Roots*. Gilbert, "Re-visioning Local Biologies," describes a productive HIV research collaboration between Senegalese and American scientists from the mid-1980s, but the exponential rise in HIV/AIDS research funding redirected efforts elsewhere; see also Crane, *Scrambling for Africa*. On the uneven landscapes created by "enclaves" of well-resourced global health research within deteriorating national scientific and health care infrastructures, see Geissler, "Archipelago of Public Health."

67   Indeed, toxicologists have suggested their work has been "doubly abandoned" by the state and by "global health." Anthropologists have similarly

observed the effects on some areas of health care and research (cancer, diabetes, hypertension, sickle-cell anemia) of neglect relative to investment in high-priority areas of global health investment such as HIV/AIDS. See Livingston, *Improvising Medicine*; Fullwiley, *The Enculturated Gene*; Mulemi, "Technologies of Hope"; Whyte, "Publics of the New Public Health"; and Whyte, "Knowing Hypertension and Diabetes."

68 Geissler, "Parasite Lost."

69 Arnaut and Blommaert, "Chthonic Science"; Tousignant, "Broken Tempos"; Osseo-Asare, "Scientific Equity"; Droney, "Ironies of Laboratory Work."

70 Waast, "L'état des sciences en Afrique"; Geissler et al., "Sustaining the Life of the Polis"; Tousignant, "Broken Tempos."

71 Pfeiffer and Chapman, "Anthropological Perspectives."

72 Ridde, "Per Diems Undermine Health"; Loewenson, "Structural Adjustment and Health"; Prince, "Situating Health and the Public"; Pfeiffer, "International NGOs"; Turshen, *Privatizing Health Services*; Masquelier, "Dispensary's Prosperous Façade"; Jaffré and Olivier de Sardan, eds., *Une médecine inhospitalière*; Whyte, "Pharmaceuticals as Folk Medicine."

73 Foley, *Your Pocket*; Livingston, *Improvising Medicine*; Wendland, *A Heart for the Journey*.

74 See "Unprotection," on the Wiktionary website, https://en.wiktionary.org/wiki/unprotection, accessed October 17, 2017.

75 See, for example, Monosson, "Chemical Mixtures."

76 Jas, "Public Health and Pesticide Regulation."

77 Boudia and Jas, "Risk and Risk Society"; Boudia and Jas, *Toxicants, Health and Regulation*; and Boudia and Jas, *Powerless Science?* This echoes the modernization theory of "treadmill of production," which, as described by Pellow, explains capitalism's continuous production of social and environmental harm in part by the state's contradictory roles "to both facilitate capital growth and provide for social welfare and environmental protection": *Resisting Global Toxics*, 21.

78 Sellers, *Hazards of the Job*, especially 139–184, describes the development of occupational hygiene science in the 1920s in university laboratories, notably at Harvard, that sought to neutralize the political tenor of prior Progressive Era assessments of workplace hazards and legitimate corporate funding for their activities. See also Vogel, "From 'The Dose' to 'The Timing.'"

79 Markowitz and Rosner, *Deceit and Denial*, 5.

80 Sellers, *Hazards of the Job*, 183. For France, Nathalie Jas describes, in "Public Health and Pesticide Regulation," a more autonomous, cautious, and critical stance on the part of academic toxicologists. Yet their involvement in regulatory activities, where they came up against those who sought to promote the use of toxic substances such as pesticides in economically productive activities, similarly resulted in the legitimization of both pesticide use and of a regulatory system they knew to be insufficiently stringent.

81 Boudia and Jas, *Toxicants Health and Regulation*; Murphy, *Sick Building Syndrome*, 88.

82 Murphy, *Sick Building Syndrome*, 91.

83 Nash, *Inescapable Ecologies*, chapter 4.

84 Murphy, *Sick Building Syndrome*.

85 Vogel, *Is It Safe?*; Langston, *Toxic Bodies*.

86 Pellow, *Resisting Global Toxics*, 29–30; Environmental Defense Fund, *Toxic Ignorance*; Elderkin, Wiles, and Campbell, *Forbidden Fruit*; Frickel and Vincent, "Hurricane Katrina."

87 Markowitz and Rosner, *Deceit and Denial*.

88 Boudia and Jas, *Powerless Science?*, 21.

89 Discussion, "Workshop on Infrastructures of Exposure: Toxicity, Temporality and Political Economies in Africa," Cambridge, UK, March 14, 2014.

90 Diop et al., "Contamination par les aflataxies"; Diop et al., "Bioaccumulation des pesticides."

91 Fortun and Fortun, "Scientific Imaginaries and Ethical Plateaus," 44.

92 Fortun and Fortun, "Scientific Imaginaries and Ethical Plateaus," 48.

93 Fortun and Fortun, "Scientific Imaginaries and Ethical Plateaus," 49.

94 Murphy, *Sick Building Syndrome*, 119–121.

95 Frickel, *Chemical Consequences*, 141.

96 Fortun and Fortun, "Scientific Imaginaries and Ethical Plateaus," 49.

97 Guyer, "Prophecy and the Near Future," 411.

98 Mbembe and Roitman, "Figures of the Subject."

99 Ferguson, *Expectations of Modernity*, 236.

100 Cooper, "Africa's Pasts"; Diouf, "Urban Youth and Senegalese Politics"; Werbner, *Memory and the Postcolony*; De Jorio, "Introduction to Special Issue"; De Jong and Rowlands, *Reclaiming Heritage*.

101 Piot, *Nostalgia for the Future*, 163–164. The reference is to Guyer, "Prophecy and the Near Future."

102 On how health care seekers navigate the fragmentation of "projectified" HIV/AIDS care, see, for example, Whyte et al., "Therapeutic Clientship." On the temporal effects of urban poverty and infrastructural absence, see De Boeck, "'Divining' the City."

103 Clarke, "A Technocratic Imperial State?"; Kusiak, "Instrumentalized Rationality"; Kusiak, "'Tubab' Technologies"; Droney, "Ironies of Laboratory Work"; Osseo-Asare, *Bitter Roots*.

104 Geissler, "Parasite Lost"; Okeke, "African Biomedical Scientists."

105 Masquelier, "Teatime."

106 Thompson, "Time." An innovative application of this idea to scientific work is Charles Thorpe's analysis of the social organization of time in atomic bomb research at the Los Alamos Laboratory during the war. Thorpe argues in "Against Time" that scheduling regulated daily work and became an end in itself, a work ethic that cut scientists off from the larger moral and social implications of bomb research.

107 Lee, "Weber, Re-enchantment and Social Futures."

108 Geissler et al., eds., *Traces of the Future*.

1 Mamadou Fall, personal communication, July 2010.

2 Rose Diene, personal communication, February 2011.

3 "Historique," Université Cheikh Anta Diop, Faculté de Médecine, de Pharmacie et d'Odontologie (FMPO), accessed August 13, 2015, http://fmpos.ucad. sn
/index.php?option=com_content&view=article&id=47&Itemid=54.

4 On ethnographic and historical approaches to scientific capacity, see Geissler and Tousignant, "Capacity as History and Horizon." My questions here are related to, but more historically oriented—that is, concerned with the past *in* the past as well as in the present—than our approach in Geissler et al., eds., *Traces of the Future* (see in particular the introduction by Geissler and Lachenal).

5 Rose Diene, personal communication, February 2011.

6 Babacar Niane, interview by author, August 19, 2010.

7 Among Gras's abandoned archives (a few boxes of files left on the shelves of an office in the toxicology lab) was this document: Coleman Instruments, "Operating Directions 50–900, Perkin-Elmer Coleman Mercury Analyzer MAS-50," Maywood, IL, 1972. Although these documents are not officially kept or classified as an archival collection, either by an individual or an institution, and I believe they have been moved, either for storage or disposal, I refer to their location as the "Gras Collection," which I found in the Laboratoire de Toxicologie et d'Hydrologie (LTH) in the Faculty of Pharmacy at Université Cheikh Anta Diop (UCAD), Dakar.

8 Gras and Mondain, "Microdosage colorimétrique," mention that the apparatus is affordable but complain that maintenance is difficult and slow to obtain in Dakar. They thus point out the need to develop cheap, accessible methodologies in "developing countries." Michele Sourdeau (Beckman Instruments International) to Georges Gras, July 16, 1980, Gras Collection, provides information about the delays and costs of obtaining maintenance services and parts for a DBT Spectrophotometer in Dakar.

9 Mondain, "Travaux pratiques de toxicologie," n.d., is a document I found in another set of "archives," piled into filing cabinets, along with student theses, in the entranceway to the analytical chemistry teaching lab. These documents seem to have been part of the LCAT's former administrative archives, but they have not been classified, nor necessarily stored as archives. I refer to this set of documents as the LCAT Unofficial Archives (which is different from the LTH Administrative Archives cited in chapter 5, which are ordered and kept by an administrator as part of the functioning of the lab's office). Although undated, the document identifies Mondain as an "assistante stagiaire," a junior faculty rank, which was her status in 1972 (by 1973 she was promoted to "assistante," according to the *Bulletins et Mémoires de la Faculté de Médecine, Pharmacie et Odonto-Stomatologie de Dakar*). Several of the

exercises described in the booklet, such as the detection of ethanol in blood and barbiturates in urine, the dosage of parathion and the determination of cholinesterase activity in blood were mentioned by Fall in 2010 as part of the current syllabus.

10 Doudou Ba, interview by author, March 29, 2010.

11 Mathilde Cabral, interview by author, September 1, 2010.

12 Affa'a and Des Lierres, *L'Afrique noire*, 39.

13 Eisemon and Salmi, "African Universities and the State."

14 "Extrait de l'Info Senegal—Bulletin Quotidien de l'Agence de Presse Senegalaise (24/25/9/61): L'important Accord Culturel France-Senegalais sur l'Université de Dakar, Témoignage de Confiance," Archives of the Faculty of Medicine, Pharmacy and Dentistry, University Cheikh Anta Diop, Dakar (hereafter: FMPOS Archives); Blum, "Sénégal 1968."

15 These were the words of its first French rector, Lucien Paye, cited in Eisemon and Salmi, "African Universities and the State," 156; Cruise O'Brien, *White Society in Black Africa*, 170–171.

16 "Extrait du Journal Officiel de la République du Sénégal, Loi n. 67–45 du 13 Juillet 1967 relative à l'Université de Dakar," FMPOS Archives.

17 Affa'a and Des Lierres, *L'Afrique noire face à sa laborieuse appropriation de l'université*, 40, and "Accord de coopération en matière d'enseignement supérieur entre le Gouvernement de la République française et le Gouvernment de la République du Sénégal (ensemble deux annexes) signé à Paris le 29 mars 1974, 1701–1707, http://www.ambasseneparis.com/index.php/cooperation.html, accessed August 13, 2015.

18 Global Pharma Health Fund E.V., "Minilab—Protection against Counterfeit Medicines," http://www.gphf.org/web/en/minilab, accessed August 7, 2014; emphasis mine.

19 The DQI was created in 2000 and was expanded and renamed in 2009 as Promoting the Quality of Medicines (PQM). See United States Pharmacopeial Convention, "Promoting the Quality of Medicines in Developing Countries," http://www.usp.org/global-health-programs/promoting-quality-medicines -pqmusaid, accessed August 13, 2015.

20 Amadou Diop, interview by author, August 25, 2010.

21 USP, "DQI Proposed Work Plan, Senegal, October 1, 2008–September 30, 2009," http://www.pmi.gov/docs/default-source/default-document-library /implementing-partner-reports/dqi-proposed-work-plan-senegal-fy2009 .pdf?sfvrsn=4, accessed August 13, 2015.

22 LeadCare II promotional website, http://www.leadcare2.com/, accessed August 13, 2015.

23 Blacksmith Institute, "Project Completion Report."

24 *Global health* is being defined here as pertaining to transnational funding based on determinations of priority made on a global rather than on a national scale. On the ways in which "the global" supersedes or bypasses national determinations of responsibility, citizenship, and sovereignty, see

Brown, Cueto, and Fee, "The World Health Organization"; Lakoff, "Two Regimes of Global Health"; Nguyen, "Government by Exception"; Redfield, "Bioexpectations"; and Rees, "Humanity/Plan."

25 Quet, "Sécurisation pharmaceutique"; Blacksmith Institute and Global Alliance on Health and Pollution, "The Poisoned Poor." These efforts to define fake medicines and poisoning as global health problems use a similar metrics, scaling, and temporal emergency by which other conditions, notably HIV/AIDS, were constituted into global health problems of which the relations of exchange, expectation, and obligation were stretched beyond national territory, allowing for claims for therapy to be addressed to entities other than the state. See especially Nguyen, "Antiretroviral Globalism"; and Whyte et al., "Therapeutic Clientship."

26 Redfield, "Bioexpectations."

27 In the years following my fieldwork, several studies were published in international journals, for example: Dieme et al., "Caractérisation physico-chimique"; Diop et al., "Étude de la contamination par les elements"; Cabral et al., "Low-level Environmental Exposure."

28 On inequality in transnational collaborations, see Crane, "Unequal 'Partners'"; Moyi Okwaro and Geissler, "In/dependent Collaborations"; and Brown, "Global Health Partnerships." On the sense of lack that results from comparisons between African and better-equipped spaces of care and research, see Wendland, "Moral Maps," and Droney, "Ironies of Laboratory Work."

29 Fieldnotes, Dakar, July 15, 2010. The original French is: "L'université; le patrimoine de l'étudiant, le socle d'une nation. Pfizer, une unité de production de médicaments au service de l'Afrique."

30 Dalberto, Charton, and Goerg, "Urban Planning."

31 The university was formed from an Institute of Higher Education, created in 1950, by merging several institutions of higher education, including the School of Medicine (the first, founded in 1918) and the IFAN (French Institute for Black Africa), founded in 1935 (see, e.g., Eisemon and Salmi, "African Universities.") The former buildings of both institutions are back to back in the downtown Plateau district; one is now a regional office of the UN development program (UNDP) and the other a museum of African art. The current location of the university is in the Fann neighborhood, close to the Medina, where colonial authorities attempted to move Africans from the Plateau following a plague outbreak in 1914; see Echenberg, *Black Death, White Medicine.*

32 Cruise O'Brien, *White Society in Black Africa.*

33 Uduku, "Modernist Architecture and 'the Tropical'"; Livsey, "Suitable Lodgings."

34 Cooper, *Decolonization and African Society*; Kusiak, "Instrumentalized Rationality."

35 Blum, "Sénégal 1968."

36 Mills, "Life on the Hill." For a similar argument on laboratory science, see Osseo-Asare, "Scientific Equity."

37 Eisemon and Salmi, "African Universities," 155; Gaillard, Hassan, and Waast, "Africa," 177.

38 Cruise O'Brien, *White Society in Black Africa*, 170–171; "Série 2S: Dossiers d'étudiants de pharmacie," FMPOS Archives.

39 Attisso, "Contrôle de la qualité," 12.

40 Attisso, "Contrôle de la qualité des médicaments au Sénégal," 13–17.

41 On plans to Senegalize pharmaceutical production and pharmacy ownership, see Tousignant, "The Qualities of Citizenship."

42 On pharmacy education and the changes in the conditions of private pharmacy ownership in Senegal, see Patterson, *Pharmacy in Senegal*.

43 Gaye, "Contribution à l'évaluation de la performance."

44 Osseo-Asare, "Scientific Equity."

45 Mamadou Fall, personal communication, July 26, 2010.

46 Félicité Bandiaky, personal communication, July 2010.

47 See also Claire Wendland's ethnography of medical education in Malawi, *A Heart for the Work*, describes the development of embodied technical skills in a resource-poor setting and especially the confrontation between theoretical skills and their application under difficult conditions of care.

48 Mamadou Fall, personal communication, July 26, 2010.

49 Goudiaby, "Le Sénégal dans son appropriation de la réforme."

50 This was particularly evident in the lectures on quality management I attended in 2010, as well as in the color pamphlets advertising the degrees posted in the hallway of the Faculty of Pharmacy. Amadou Diouf, the head of the university lab, who spearheaded these programs, also initiated quality-management procedures at the National Drug Control Laboratory (Laboratoire de contrôle des médicaments) and the Poison Control Centre (Centre Anti-Poison), both Ministry of Health institutions that he directed. Thesis topics also show that lab members encourage students to work on topics of interest to the private sector, sometimes in collaboration with former members of the lab or other personal connections who work in private pharmacy and quality-management consulting.

51 In a 2008 focus section in *Isis* on "laboratory history," Robert Kohler and Thomas Gieryn reflect on the ways in which modern laboratories achieve "placelessness" as a marker of the mobility and universality of the knowledge they produce. In the same section, Graeme Gooday instead argues that a large proportion of laboratories, including regulatory testing and teaching laboratories, actively sought to "place" themselves in specific, located networks to validate and legitimate the knowledge they produced. See Kohler, "Lab History;" Gooday, "Placing or Replacing the Laboratory?"; and Gieryn, "Laboratory Design."

52 Anderson, "Frozen Archive"; Kowal and Radin, "Indigenous Biospecimen Collections."

53 Shapin, *Scientific Life*.

54　The value of the West African CFA franc currency was halved in 1994 as a measure imposed by international financial institutions to stimulate exports.

55　Mathilde Cabral, interview by author, August 25, 2010.

56　Mathilde Cabral, interview by author, September 1, 2010.

57　Amadou Diop, interview by author, August 25, 2010.

58　DeSilvey and Edensor, "Reckoning with Ruins."

59　See, for example, González-Ruibal, "Dream of Reason"; Tousignant, "Half-built Ruins."

60　DeSilvey and Edensor, "Reckoning with Ruins," 468; Edensor, "Ghosts of Industrial Ruins."

61　Stoler, "Imperial Debris," 194.

62　Stoler, "Imperial Debris," 197 and 204.

63　DeSilvey and Edensor, "Reckoning with Ruins," 472; Stoler, "Imperial Debris," 202.

64　DeSilvey and Edensor, "Reckoning with Ruins," 467; Graham and Thrift, "Out of Order," 5–6.

65　Schaffer, "Easily Cracked."

66　Boudia and Soubiran, "Scientists and Their Cultural Heritage."

67　Jardine and Wilson, "Recent Material Heritage"; Boudia and Soubiran, "Scientists and Their Cultural Heritage."

68　Masquelier, "Behind the Dispensary's Prosperous Facade"; De Boeck, "Postcolonialism, Power and Identity."

69　Stoler, "Imperial Debris," 196.

70　Stoler, *Along the Archival Grain*, 20.

71　Derrida and Prenowitz, "Archive Fever," 10. See also Povinelli, "Woman on the Other Side."

72　Cited in Anderson, "Frozen Archive," 5.

73　Povinelli, "Woman on the Other Side," 152, writes: "But if 'archive' is the name we give to the power to make and command what took place here or there, in this or that place, and thus what has an authoritative place in the contemporary organization of social life, the postcolonial archive cannot be merely a collection of new artifacts reflecting a different, subjugated history. Instead, the postcolonial archive must directly address the problem of the endurance of the otherwise." In other words, the postcolonial archive must take a new form and domicile in order to accommodate nondominant ways of knowing and being other than in their relation to domination. See also Isaacman, Lalu, and Nygren, "Digitization"; and Harris, "The Archival Sliver."

74　Gaillard, "Senegalese Scientific Community," reports that foreign sources accounted for 75 percent of research funding in 1972, dropping to 67 percent in 1975 and 68 percent in 1986. The proportion of French foreign aid dropped significantly in the 1980s but was partly replaced by other sources, notably by World Bank funding.

75 Geissler, "Archipelago of Public Health." See also Prince, "Situating Health and the Public," and Geissler et al., "Sustaining the Life of the Polis."

76 Geissler, "Parasite Lost"; Waast, "L'état des sciences," 5; Gaillard et al., "Africa," 177.

78 Gaillard and Waast, in "La recherche scientifique en Afrique," 14, describe how scarce public funds are consumed by salaries, while the costs of reagents, fieldwork, and dissemination are covered only sporadically, depending on precarious influxes of national public or foreign funding. This results in the interruption of research and the nonpublication of results, aggravated by the adoption of short-term scientific planning, "without follow-up and without strategy."

79 Geissler et al., "Sustaining the Life of the Polis."

80 Livsey, "'Suitable Lodgings'"; Osseo-Asare, "Scientific Equity."

81 Moyi Okwaro and Geissler, "In/dependent Collaborations"; Fullwiley, *Enculturated Gene*; Crane, *Scrambling for Africa*.

82 Wendland, *A Heart for the Work*, 179; Livingston, *Improvising Medicine*, especially 93–118, on nursing and the political promise of care.

83 Wendland, *A Heart for the Work*, 137.

84 On the iatrogenic harm of human/material incapacity, see Julie Livingston, commentary in "Roundtable: Capacity as Moral Imperative," and Marissa Mika, "Cobalt Blues: Radiotherapy Technopolitics in Uganda," oral presentation, Making Scientific Capacity in Africa Conference, University of Cambridge/CRASSH, June 13–14, 2014.

2 · ADVANCEMENT

1 Pille, "Leçon inaugurale."

2 Osseo-Asare, "Bioprospecting and Resistance"; Hokkanen, "Imperial Networks"; Kerharo and Bouquet, "Plantes médicinales et toxiques." Arrow and fishing poisons are generally used to stun animals to catch them for human consumption. Ordeal poisons are used in witchcraft or sorcery trials to determine the guilt of the accused, as indicated by vulnerability to the poison.

3 Pille, "Lecon inaugurale," 7; emphasis mine.

4 Babacar Niane, interview by author, August 19, 2010.

5 Babacar Niane, interview by author, August 19, 2010; emphasis mine.

6 Babacar Niane, interview by author, March 30, 2010.

7 Babacar Niane, interview by author, August 19, 2010.

8 See, for example, Geissler, "Parasite Lost"; Geissler, "What Future Remains?"; and Droney, "Ironies of Laboratory Work."

9 Hecht, "Rupture-Talk in the Nuclear Age," 693.

10 Lachenal, "Intimate Rules of French Coopération."

11 See, for example, Reid-Henry, *The Cuban Cure*; Pollock, "Places of Pharmaceutical Knowledge-Making."

12 Bonneuil and Petitjean, "Les chemins de la création de l'ORSTOM"; Adas, "Scientific Standards and Colonial Education."

13 Livingston, *Improvising Medicine*, 39. See also Mika, "Fifty Years of Creativity."

14 Medical schools were among the first institutions of higher education in the colonies. In French West Africa, a medical school was created in Dakar in 1918, but its main aim was to create a corps of trained auxiliaries. On the aims and limits of colonial higher education, see Sabatier, "'Elite' Education in French West Africa." Bonneuil and Petitjean, in "Les chemins de la création de l'ORSTOM," 148, note that prior to 1945, there was little concern among policymakers with the training of scientists within colonized populations.

15 There is a large body of literature on the training and professionalization of African medical personnel emphasizing racism and gendering in the definition of professional roles and mobility. See Iliffe, *East African Doctors*; Patterson, "Women Pharmacists in Twentieth-Century Senegal"; Patterson, *Pharmacy in Senegal*; Barthélémy, "La professionalisation des africaines en AOF (1920–1960)"; and Marks, *Divided Sisterhood*. For a good historical overview, see Wendland, *A Heart for the Work*.

16 Bonneuil and Petitjean, in "Les chemins de la création de l'ORSTOM," 141, write of "the strong centralizing pressure of the Parisian mandarins, who repeatedly called for a division of labor between, in the colonies, observation and collection and, in the metropole, the in-depth study of data and materials, and synthesis." See also Hountondji, "Scientific Dependence in Africa Today."

17 Bonneuil and Petitjean, "Les chemins de la création de l'ORSTOM," 120.

18 On the laboratory metaphor of Africa as a vast field of research but also on how field sciences in Africa fueled new understandings of health and ecology, see Tilley, *Africa as a Living Laboratory*.

19 Hountondji, in "Scientific Dependence in Africa Today," 7–9, argues that this subordination to metropolitan infrastructure made colonial science an "impoverished science."

20 Bonneuil and Petitjean, "Les chemins de la création de l'ORSTOM," 141–142. On the mobilization of technocratic expertise, see Cooper, *Decolonization and African Society*, and Clarke, "A Technocratic Imperial State?"

21 "Reclassement des médecins, pharmaciens et sages-femmes africains, 1.1.1945," *Journal Officiel de l'Afrique Occidentale Française* 2171 (1945): 529; "Décision n° 416 portant autorisation d'un ex-médecin africain à exercer en pratique privée à Dakar, 25.1.1956," *Journal Officiel du Sénégal* 3014 (1956): 111.

22 "Arrêté 2220 portant mise au concours de bourses d'études pour 1949 entre les médecins africains en service en AOF, AEF, Cameroun et Togo, pour poursuivre leurs études en vue d'accéder au diplôme d'État de docteur en médecine, 29.4.1949," *Journal Officiel de l'Afrique Occidentale Française* 2414 (1949): 579.

23 Chabas, "L'Institut des Hautes Etudes de Dakar"; Paye, "Training Administrative Staff."

24 Kusiak, "'Tubab' Technologies," 227.

25 Cruise O'Brien, *White Society in Black Africa*, 165 and 188.

26 Cruise O'Brien, *White Society in Black Africa*, 170; Kusiak, "'Tubab' Technologies"; Lachenal, "Intimate Rules of French Coopération."

27 Hecht, "Rupture-Talk in the Nuclear Age," 693; Lachenal, "Intimate Rules of French Coopération," 386; Kusiak, "'Tubab' Technologies," 227.

28 Hecht, "Rupture-Talk in the Nuclear Age"; Lachenal, "Intimate Rules of French Coopération"; Kusiak, "'Tubab' Technologies." For a different discussion of how race is both erased by and foundational to development ideology, see Soske, "'The Dissimulation of Race.'"

29 Lachenal, "Intimate Rules of French Coopération," 377.

30 Cruise O'Brien, *White Society in Black Africa*.

31 Cruise O'Brien, *White Society in Black Africa*, 166 and 173.

32 On Négritude as a political philosophy and citizenship project, see Markovitz, *Léopold Sédar Senghor*. On the aesthetic dimensions of Négritude, see Harney, *In Senghor's Shadow*. Harney emphasizes the syncretic dimensions of Négritude, as does Kilroy-Marac in "Impossible Inheritance." For a good overview of the history and uses of Négritude, see Vaillant, "Perspectives on Leopold Senghor." Among the critics of Négritude and Senghor's positions on decolonization were the Cameroonian philosopher Marcel Towa and Frantz Fanon. See Wilder, *Freedom Time*.

33 On Senghor's (and Aimé Césaire's) claims to both "Frenchness" and "Western civilization" as shared historical heritage and as driving their cosmopolitan aspirations, see Wilder, *Freedom Time*.

34 Bernabé, "Senghor."

35 Speech given at the University of Dakar cited in Niang, "De la politique scientifique senghorienne."

36 Abraham, "Landscape and Postcolonial Science," 170. Abraham and Gyan Prakash explore the role of science in Indian nationalism, foregrounding how elite nationalist discourse sought both to depart from the coercive colonial uses of science and to appropriate its power to legitimize an authentic and modern postcolonial nation; see Abraham, "Contradictory Spaces"; Abraham, "Science and Power in the Postcolonial State"; and Prakash, *Another Reason*. On Nkrumah's science policies, see Droney, "Ironies of Laboratory Work"; Osseo-Asare, "Scientific Equity"; and Osseo-Asare, *Bitter Roots*.

37 Arnold, "Nehruvian Science and Postcolonial India."

38 Diouf, *L'historiographie indienne en débat*, 5–9.

39 Diop, *Sciences et philosophies*, 53–72 and 123–131. The text quoted is on p. 128 and is drawn from the opening keynote for the ninth biannual conference of the West African Scientific Society at the University of Dakar in March 1974.

40 Diop, *Sciences et philosophies*, 128–129.

41 As Frantz Fanon's more radical critique of the legacy of colonization and of Senghor's moderate and assimilationist position as that of a "mimic man" would suggest, see Wilder, *Freedom Time*, Kindle edition location 3324.

42　Arnold, "Nehruvian Science and Postcolonial India," distinguishes between "science as authority" and "science as delivery." The former would apply here.

43　Harney, *In Senghor's Shadow*.

44　Diouf, "Le clientélisme, la 'technocratie' et après?," 250 and 257; Tourte and Le Moigne, "L'équipement rural au Sénégal"; Schumacher, *Politics, Bureaucracy, and Rural Development in Senegal*.

45　Kusiak, "'Tubab' Technologies"; Kusiak, "Instrumentalized Rationality."

46　Kusiak, "'Tubab' Technologies"; Diouf, "Le clientélisme, la 'technocratie' et après?"; Markovitz, in *Léopold Sédar Senghor*, refers to Senghor's "politics of technicity."

47　Abraham calls for case studies of science as practice. For a more nuanced understanding of how place and power inflects science in postcolonial locations, see his "Contradictory Spaces."

48　On pharmaceutical production, see Tousignant, "Qualities of Citizenship," 100. For an extensive overview of the Senegalese economy and the emergence of public-private industries in the 1970s, see Rocheteau et al., *Pouvoir financier et indépendance économique en Afrique*.

49　Cruise O'Brien, *White Society in Black Africa*.

50　République du Sénégal, "Quatre années de fonctionnement," 1.

51　République du Sénégal, "Quatre années de fonctionnement," 5.

52　République du Sénégal, "Quatre années de fonctionnement," 75.

53　République du Sénégal, "Quatre années de fonctionnement," 75, states that 661 people had at least a partial research role (although it was noted that academic lecturer-researchers could spend no time on research activities), of which 480 were employed by the university. Thus, the "human potential" of the university for research was deemed "significant but underutilized." Out of these, "198.3" were Senegalese, of which "166.5" were employed by the university.

54　République du Sénégal, "Quatre années de fonctionnement," 76.

55　"Série 2S: Dossiers d'étudiants de pharmacie," FMPOS Archives. This folder contains thirty-seven student files from the 1960s. These files were probably not archived systematically (there is no evidence that they were) and cannot be taken as representative of all students during this period. Yet given the relatively small number of pharmacy students during this period, this sample is probably indicative of broader trends. Files contain information about the nationality of students, often about the occupation of parents, and include requests for transfers. Of the thirty-seven archived files, sixteen are of French students and ten are of non-Senegalese Africans. Most French students had fathers who worked as technical assistants in Senegal (usually in Dakar). Fifteen files suggest that students who had begun pharmacy studies in Dakar then transferred to France (this includes non-French students but not transfers to other faculties of the University of Dakar). However, very few archived files indicate completed courses of study, which may indicate a bias in the archiving of interrupted studies. For limitations on terms of service for technical assistants, see Cruise O'Brien, *White Society in Black Africa*.

56 Including in toxicology, as a professor in legal and occupational medicine did: Ceccaldi, "Evolution de la toxicologie contemporaine."

57 Blanc and Jaussaud, in "Les leçons inaugurales," write, on the tradition of inaugural lessons, that new holders of existing chairs tended to recapitulate its history, while new chairholders generally justified its institutional existence.

58 Monnais and Tousignant, "Values of Versatility."

59 Pille, "Leçon inaugurale," 3.

60 Pille, "Conditions particulières à l'expertise toxicologique tropicale."

61 On the history of toxicology in France, see Irissou, "Brève histoire de la toxicologie," and Le Moan, "L'enseignement de la toxicologie." On the versatility of French pharmacists, see, for example, Berman, "J. B. A. Chevallier."

62 French colonial pharmacists were military officers who performed a variety of functions, including running government supply pharmacies, hospital labs, and control labs, and conducting research and lecturing in medical schools, often concurrently and across the disciplines of chemistry, bacteriology, toxicology, biochemistry, pharmacology, and botany. See Monnais and Tousignant, "Values of Versatility."

63 For example, the medicinal plant survey mission entrusted by the governor general of French West Africa to the colonial pharmacist Numa Laffitte in 1935, or the survey mission for plant-based fuel and alcohol substitutes in Niger entrusted by the governor general of French Central Africa to the colonial pharmacist Joseph Kerharo. See Laffitte, *La pharmacopée indigene*, and "Dossier Biographique: Joseph Kerharo," Bibliothèque Interuniversitaire de Pharmacie, Faculty of Pharmacy, Paris, France.

64 Pille, "Leçon inaugurale," 7.

65 Osseo-Assare, "Bioprospecting and Resistance."

66 Pille and Palancarde, "Quelques aspects actuels de l'expertise toxicologique."

67 Monnais, "Des poisons qui en disent long."

68 Especially Pille, "Conditions particulières à l'expertise toxicologique tropicale."

69 Jas, "Public Health and Pesticide Regulation"; Le Moan, "L'enseignement de la toxicologie."

70 "Extrait du Journal Officiel de la République Française du 29 mai 1960, p. 4874: Décret du 23 mai 1960 portant transformation de l'École Nationale de Médecine et de Pharmacie de Dakar en Faculté Mixte de Médecine et de Pharmacie," FMPOS Archives.

71 Although there is a precedent in Emile Perrot's project of exploiting colonial raw materials to advance both French industry and French plant chemistry, in resistance to the ascendance of synthetic chemistry, see Debue-Barazer, "Des simples aux plantes médicinales."

72 Pille probably died in 1964.

73 Pille and Palancarde, "Quelques aspects actuels de l'expertise toxicologique."

On expatriate suicide, see also Collomb et al., "Les conduites suicidaires à Dakar," and Cruise O'Brien, *White Society in Black Africa*, 229.

74 Pille, "Conditions particulières à l'expertise toxicologique tropicale."

75 Payet et al., "A propos d'une intoxication volontaire."

76 Payet et al., "Le contexte humoral lipidique."

77 Pille et al., "Quelques aspects de la recherche chimique."

78 Sankalé et al., eds., *Dakar en devenir*, is illustrative of this trend.

79 Livingston, *Improvising Medicine*, 39.

80 Césaire et al., "Les intoxications au Sénégal."

81 Gras, "La lutte chimique contre les moustiques." This doctoral thesis was about "problems"—presumably the unintended effects on human health and the environment—of insect control in the south of France.

82 Previously, Pille and Césaire had published mainly in West African medical journals. Césaire's series of papers describing a range of uses for the apparatus he developed a decade earlier (which had at the time been reported in a French biology and chemistry journal) were published in the faculty's bulletins. After Gras, during the 1980s and 1990s, the publication rate declined overall, and most studies conducted in the lab were published in *Dakar Médical* (a local medical journal); the few international publications were based on doctoral work conducted overseas.

83 Gras and Fauran, "Dosage colorimétrique du Diacétate de Plomb Dibutyle"; Gras and Fauran, "Etude de la réaction de complexation du DPD"; Fauran and Gras, "Etude spectrophotometrique des reactions de complexation des DPD."

84 Gras et al., "Toxicologie d'un nouvel anthelminthique."

85 Fauran et al., "Dosage colorimétrique du Chloralose Alpha"; Fauran et al., "Dosage de L'Alpha Chloralose"; Fauran et al., "Métabolisme du Chloralose Alpha I"; Fauran et al., "Métabolisme Du Chloralose Alpha II"; Pellissier, "Contribution à l'étude de la toxicologie analytique du Chloralose Alpha"; Gras et al., "Toxicologie analytique Chloralose Alpha."

86 Fauran et al., "A propos d'une intoxication mortelle."

87 For example, Gras and Mondain, "Etude experimentale de la bioaccumulation," and Gras and Mondain, "Microdosage colorimétrique du mercure total."

88 For example, Gras and Mondain, "Teneur en mercure de quelques espèces de poisons"; Gras and Mondain, "Influence de la consommation de poisson"; and Gras and Mondain, "Rapport methylmercure/mercure total." They did, however, publish an article on the risks of mercury in cosmetics (skin-lighteners) in Senegal: Gras and Mondain, "Problème posé par l'utilisation des cosmétiques."

89 Georges Gras to M. Sene, directeur de la Commission de Recherche, Dakar, June 18, 1976, Gras Collection.

90 Jackson, "FAO Quelea Research in Africa."

91 Georges Gras à M. Blanc, conseiller technique, Direction des pêches mari-

times, Dakar, June 17, 1976, Gras Collection. In exchange for free fish samples, Gras proposed his lab help repair and calibrate the fishery service's MAS (Mercury Analyzer System) 50 and train a technician to perform mercury analyses at the port. These propositions were accepted (Georges Gras, "Résumé d'entretien de Monsieur Gras avec le directeur de l'Institut des Pêches," June 24, 1976, Gras Collection). A few years later, Gras's opinion on the need for food standards to increase the competitiveness of Senegalese exports was solicited by the Institut sénégalais de normalisation (Senegalese Standardization Institute), indicating the Senegalese government's interest in quality control as an economic strategy.

92   Jensen and Odhiambo, "Mission Report on Problems, Interests and Capabilities for the Training and Research in Environmental Chemistry in Selected African Countries, 24 Nov–18 Dec 1976," UNESCO, 1976: 4, Gras Collection.

93   Kouthon, "Rapport d'une mission au Sénégal, 18–24 janvier 1979 puis 30 janvier–2 février 1979," Rome: FAO, 1979: 3–4, Gras Collection.

94   Cailleux, "Identification rapide des formes médicamenteuses."

95   Broadhead, "Officer Ugg, Mr Yuk, Uncle Barf"; Rambourg-Schepens et al., "Rapport de la mission InVS/Afsse."

96   These obstacles are noted in Kouthon, "Rapport d'une mission," and Jensen and Odhiambo, "Mission Report on Problems." The latter also note the constraints posed by lab members' training as pharmacists rather than as "true research chemists."

97   For a similar argument about a road, see Redfield, "Half-Life of Empire."

98   According to available faculty bulletins, Ciss took up his post in Dakar in 1979 or 1980, while Gras is listed as staff until 1982. They may have overlapped more than Ciss suggests, or one of them may not have been physically present during this time. A letter to Gras in 1978 indicates that he may have begun his post in Montpellier earlier but kept the affiliation in Dakar in order to continue supervising his students.

99   This section draws on four interviews I conducted with Niane in his home and his retail pharmacy in March and August of 2010, as well as on the CV in his possession that he allowed me to make a copy of.

100 M. Crame, "Présentation," March 1, 1959, accessed July 23, 2011, http://www.ltidelafosse.net/.

101 The other sixteen were not necessarily French; many came from other French-speaking African countries to study in Dakar.

102 Shapin, "Invisible Technician."

103 Doudou Ba, interview by author, March 29, 2010; Mounirou Ciss, interview by author, February 22, 2010.

104 Cooper, "Conflict and Connection," 1519.

105 Timmermans, "A Black Technician and Blue Babies."

106 Timmermans, "A Black Technician and Blue Babies," 222.

107 Abraham, "Contradictory Spaces," 217.

108 This information is derived from the *Bulletins de la Faculté de Médecine, Pharmacie et d'Odonto-Stomatologie de Dakar*, published as a serial.

109 Doudou Ba, interview by author, March 29, 2010.

3 · ROUTINE RHYTHMS AND THE REGULATORY IMAGINATION

1 Exposé des travaux scientifiques de Monsieur Doudou Ba, maitre assistant de Chimie Analytique a la Faculte de Medecine et de Pharmacie de Dakar, Senegal, unpublished manuscript, LCAT Unofficial Archives; Mounirou Ciss, interview by author, February 22, 2010.

2 Mounirou Ciss, interview by author, February 22, 2010.

3 Babacar Niane, interview by author, August 5, 2010.

4 Babacar Niane, interview by author, August 5, 2010.

5 Diouf, "Le clientélisme, la 'technocratie' et après?"; Diouf, "Senegalese Development."

6 Bromley, "Making Sense of Structural Adjustment."

7 Diouf, "Le clientélisme, la 'technocratie' et après?"

8 Bromley, "Making Sense of Structural Adjustment."

9 Piot, *Nostalgia for the Future*, 66.

10 Mbembe and Roitman, "Figures of the Subject"; Ferguson, *Expectations of Modernity*; Geissler, "Parasite Lost."

11 Riddell, "Things Fall Apart Again."

12 Diouf, "Fresques murales et écriture de l'histoire."

13 Haddar, *Séminaire sur les conséquences sociales*. This collection, like many others analyzing the social impact of adjustment measures, focuses especially on the health and education sectors. See also Pfeiffer and Chapman, "Anthropological Perspectives."

14 Riddell, "Things Fall Apart Again"; Haddar, *Séminaire sur les conséquences sociales*; Pfeiffer and Chapman, "Anthropological Perspectives."

15 See, for example, Whyte, "Pharmaceuticals as Folk Medicine"; Banégas and Warnier, "Nouvelles figures de la réussite"; and Geissler et al., "Sustaining the Life of the Polis."

16 Diop, "Essai sur 'l'art de gouverner.'"

17 Cited by Diop, "Essai sur 'l'art de gouverner,'" 15.

18 Diouf, "Fresques murales et écriture de l'histoire"; Haddar, *Séminaire sur les conséquences sociales*.

19 Fassin, "Du clandestin à l'officieux"; Fassin, "Vente illicite des médicaments." See also Fall et al., "Gouvernance et corruption."

20 Tousignant, "Qualities of Citizenship."

21 Jaffré and Olivier de Sardan, eds., *Une médecine inhospitalière*, 148, distinguish three historical periods: 1) free health care with medicines, 2) free health care without medicines, and 3) health care and medicines for a fee.

22 Foley, *Your Pocket Is What Cures You.*

23 Fullwiley, *Enculturated Gene.*

24 See also Prince, "Situating Health and the Public"; Livingston, *Improvising Medicine*; Mulemi, "Technologies of Hope."

25 Whyte, "Pharmaceuticals as Folk Medicine"; Pfeiffer and Chapman, "Anthropological Perspectives"; Turshen, *Privatizing Health Services.*

26 Fall et al., "Gouvernance et corruption"; Jaffré and Olivier de Sardan, eds., *Une médecine inhospitalière*; Ndoye, *La société sénégalaise face au paludisme*; Masquelier, "Behind the Dispensary's Prosperous Façade."

27 Livingston, *Improvising Medicine*, and Wendland, *A Heart for the Work*, both describe how an understanding of health workers as embodying a civic ideal of service is upheld despite material scarcity. Nostalgia for a prior era of health institutions as vectors of African modernity can be another expression of the loss of an ethos of collective service and improvement; see Geissler, "Parasite Lost," and Kilroy-Marac, "Nostalgic for Modernity."

28 Gaillard and Waast, in "L'aide à la recherche en Afrique subsaharienne," 82–83, note how national research institutions had to be restructured and/or reoriented to benefit from foreign funding sources. Gaillard, in "Senegalese Scientific Community," notes that external aid accounted for 75 percent of the national research budget in Senegal from the late 1980s.

29 Waast, "L'état des sciences," 12.

30 Gaillard, "Senegalese Scientific Community," 165.

31 Gaillard et al., "Africa," 184–185. See also Eisemon and Davis, "Quality of Scientific Training and Research."

32 Gaillard et al., "Africa," 177.

33 Waast, "L'état des sciences," 5.

34 Waast, "L'état des sciences," 5.

35 A good example is the Senegalese virologist Souleymane Mboup and his public hospital laboratory; see Gilbert, "Re-visioning Local Biologies."

36 Gaillard et al., "Africa," 185–187. On the ways in which an older national and public ethos of science remains present in emerging institutions and forms of scientific work, see, for example, Geissler, "What Future Remains?"

37 Geissler, ed., *Para-States and Medical Science.*

38 Mamadou Fall, interviewed by author, August 20, 2015.

39 Amadou Diop, interviewed by author, August 25, 2010.

40 Mamadou Fall, interviewed by author, August 20, 2015.

41 "Curriculum Vitae, Babacar Niane," personal document in possession of the author.

42 Dzidzornu, "Marine Pollution Control."

43 UNEP, "West and Central African Action Plan," 49.

44 Dzidzornu, "Marine Pollution Control," 457 and 472.

45 UNEP, "First Workshop of Participants."

46 There is a large body of literature on the politics and social effects of standards, including on the "appropriate" use of technologies. For a useful review, see Timmermans and Epstein, "World of Standards."

47   UNEP, "West and Central African Action Plan," 7; Dzidzornu, "Marine Pollution Control," 475 and 481.
48   Portmann, *State of the Marine Environment*, 5.
49   Ba et al., "Recherche et dosage des traces de métaux lourds."
50   Doudou Ba, interview by author, March 29, 2010.
51   Doudou Ba, interview by author, March 29, 2010.
52   Ndiaye, "Contrôle de qualité."
53   Monote, "Contribution à la détermination de la valeur."
54   Samou, "Les aliments enrichis de vitamine D."
55   Kane, "Contribution à l'état de la qualité," 1.
56   Bara Ndiaye, interview by author, March 8, 2010.
57   See also Tousignant, "Pharmacy, Money and Public Health."
58   Only a small fraction of thesis research involved laboratory analysis, however. A few theses were bibliographic essays on specific toxic risks or analytical methods, while the majority involved "paper-based" research, that is, based on the collection and analysis of existing data (e.g., information in clinical registers) or of responses to questionnaires (e.g., on knowledge and practices pertaining to toxic risks such as pesticide use among farmers).
59   Babacar Niane, interview by author, August 19, 2010.
60   Ndour, "Contrôle de certains métaux lourds"; Sy, "Contribution à l'étude du thé vert"; Sarr, "Contribution à l'étude de la pollution atmosphérique." A 1993 thesis on the presence of mercury in skin lighteners notes that it was not possible to use the AAS for analyses (a less sensitive colorimetric method was used instead); either the machine had broken down or the lab no longer had the necessary consumables (gas, standards, reagents) to use it.
61   Dia, "Recherche et dosage des aflatoxines"; Bathily, "Aflatoxines dans les aliments"; Sall, "Recherche et dosage des aflatoxines"; Diouf, "Aflatoxines dans l'arachide"; Kandji, "Etude de la composition chimique et de la qualité"; Saine, "Aflatoxines dans l'huile."
62   Diouf, "Aflatoxines dans l'arachide," 245.
63   Diop, "Etude de la qualité de l'eau"; Diatta, "Contribution à l'étude de la qualité des corps gras alimentaires"; Coly, "Recherche et dosage des résidus de pesticides"; Kane, "Contribution à l'étude de la qualité des beurres."
64   Niane, "Surveillance de la pollution marine."
65   I do not have the specific reference for this document, but Niane described the research to me in several of our interviews.
66   Babacar Niane, interview by author, August 19, 2010.
67   Lab members published about ten articles in *Dakar Médical* reporting results of analyses (with the AAS, GC, and HPLC) of heavy metals, pesticides, and aflatoxins between 1992 and 2004. Two others, conducted in collaboration with other laboratories in the faculty, one reporting animal toxicity tests of a medicinal plant and the other on prescribing patterns, were also published in this journal. During this period, I could find only two articles published in European journals reporting analyses performed in the Dakar lab (i.e., ex-

cluding analyses performed by its members overseas): Niane et al., "Surveillance de la pollution par les métaux lourds," and Diop et al., "Contamination par les aflatoxines."

68   Tousignant, "Pharmacy, Money and Public Health."

69   That is, employees of sanitary or commercial control services, as opposed to university staff members, who are also paid by the state.

70   Dia, "Recherche et dosage des aflatoxines," 532; emphasis mine.

71   Mounirou Ciss, interview by author, February 22, 2010.

72   Doudou Ba, interview by author, March 29, 2010.

73   Mamadou Fall, interview by author, February 4, 2010.

74   Masquelier, "Behind the Dispensary's Prosperous Façade," 271; Piot, *Nostalgia for the Future*, 38.

75   Pfeiffer and Chapman, "Anthropological Perspectives"; Foley, *Your Pocket Is What Cures You.*

76   Ba et al., in "L'arachide," 176, describe the process, which uses pressurized gaseous ammonia, and relate its use to European regulations.

77   Diouf et al., "Intoxication massive à l'ammoniac."

78   On the ways in which European food quality standards affect export agriculture in Burkina Faso and Zambia, see Freidberg, *French Beans and Food Scares.*

79   Piot, *Nostalgia for the Future*, 40. See also Bayart et al., *Criminalization of the State.*

80   Masquelier, "Behind the Dispensary's Prosperous Façade," 287.

81   Livingston, *Improvising Medicine*, 29–51.

82   For example, Amadou Diouf made this type of comment in the public defense of a thesis on pesticide residues in dried fish and in a presentation at a session on the topic of pesticides at the World Social Forum, which took place in Dakar in February 2011, as well as during CAP staff meetings when a case of collective poisoning that might have involved pesticide-contaminated lettuce was brought up.

4 · PROLONGING PROJECT LOCUSTOX,
INFRASTRUCTURING SAHELIAN ECOTOXICOLOGY

1   Gaillard and Waast, "L'aide à la recherche en Afrique."

2   FAO, *Rapport de la vingt-neuvième session.*

3   Hough, "Poisons in the System"; USAID, *Review of Environmental*; Grant, "Appendix 15," 162, describes a climate of concern about the FAO's environmental policies regarding assistance involving pesticide provision and use, and a lack of information about the risks posed by pesticides in tropical climates.

4   Galt, "Beyond the Circle of Poison."

5   Weir and Schapiro, *Circle of Poison.*

6   Weir, *Bhopal Syndrome*.

7   Bull, *A Growing Problem*.

8   Edwards and Huddleston, *Efficacy and Environmental Effects*, 1.

9   TAMS and CICP, *Locust and Grasshopper Control*.

10 USAID, *Review of Environmental Concerns*.

11 Grant, "Annex 15," 162–166.

12 FAO, *Desert Locust Research and Development Register*, 29–30.

13 Everts, *Environmental Effects of Chemical Locust and Grasshopper Control*.

14 Everts, *Environmental Effects of Chemical Locust and Grasshopper Control*, 15.

15 Van Dyk and Pletschke, "Review on the Use of Enzymes."

16 Everts, *Environmental Effects of Chemical Locust and Grasshopper Control*, 1.

17 Everts, *Environmental Effects of Chemical Locust and Grasshopper Control*, 15.

18 Everts, *Environmental Effects of Chemical Locust and Grasshopper Control*, 12.

19 Everts, *Environmental Effects of Chemical Locust and Grasshopper Control*, 1.

20 In the 1960s and 1970s, OCLALAV took an active role in research on and monitoring of pests, and then in the application of pesticides. In 1983, the French government stopped its financial and technical support to the organization. By 1988, in the middle of a major locust invasion, OCLALAV began transferring its responsibilities to the national plant protection divisions and eventually ended its activities. See Sidia et al., "OCLALAV and Its Environment"; Roy, *Histoire d'un siècle de lutte anti-acridienne*, 266–268.

21 Everts, *Environmental Effects of Chemical Locust and Grasshopper Control*, preface.

22 Everts, et al., *Locustox Project*.

23 Member states of CILSS are Burkina Faso, Cape Verde, Senegal, Mali, the Gambia, Guinea, Guinea-Bissau, Mauritania, Niger, and Chad. It became involved in pest management following the droughts and crop pest incursions of the 1970s and developed a common regulation for the registration of pesticides from the early 1990s, leading to the creation of the Sahelian Pesticides Committee in 1994: Diarra, "Pesticides Registration and Regulation," 25.

24 Meerman, "Mission au Projet FAO."

25 Ndour, "Participation aux opérations," 6.

26 Ndour, "Participation aux opérations"; Danfa and van der Valk, "Toxicity Tests with Fenitrothion."

27 Ndour, "Participation aux opérations."

28 Van der Stoep, "Development of Laboratory Toxicity Tests."

29 Van der Stoep, "Development of Laboratory Toxicity Tests," 2; van der Valk, "A Laboratory Toxicity Test."

30 Lahr et al., *Effects of Experimental Locust Control*.

31 Lahr et al., *Effects of Experimental Locust Control*.

32 Meerman, "Mission au Projet FAO"; Danfa et al., "Tests de toxicité aigue sur un parasitoide."

33 Niassy et al., "Impact of Fenitrothion Applications."

34 Van der Valk and Kamara, "Effect of Fenitrothion and Diflubenzuron," 39.

35 Meerman, "Mission au Projet FAO."

36 Thiam, "Elaboration de la table de survie partielle."

37 Lahr et al., *Effects of Experimental Locust Control.*

38 Beye et al., "Effets du fenitrothion sur les coléopteres épigés."

39 Pesticides Referee Group, "Evaluation of Field Trial Data."

40 This amount was supplemented by 202,059 USD from the Senegalese government in national staff salaries and some equipment and pesticides.

41 FAO, *Effets sur l'environnement de la lutte antiacridienne.*

42 Meerman, "Mission au Projet FAO," 4.

43 Meerman, "Mission au Projet FAO," 2.

44 Meerman, "Mission au Projet FAO," 2.

45 FAO, *Effets sur l'environnement de la lutte antiacridienne,* 20.

46 FAO, *Effets sur l'environnement de la lutte antiacridienne.*

47 MGP-Afrique, "Étude de marché."

48 FAO, "Thirty-Second Session."

49 FAO, "Thirty-Second Session"; emphasis mine.

50 Fieldnotes, Dakar, March 10, 2010.

51 Tousignant, "Insects-as-Infrastructure."

52 See, for example, O'Neill and Rodgers, eds., Special Issue on Infrastructural Violence, and Von Schnitzler, "Citizenship Prepaid."

53 Simone, "People as Infrastructure"; De Boeck, "Infrastructure."

54 Robbins, "Smell of Infrastructure."

55 Amin, "Telescopic Urbanism"; Amin, "Surviving the Turbulent Future."

56 Harvey and Knox, "Enchantments of Infrastructure."

57 Everts, "Ecotoxicology for Risk Assessment in Arid Zones," 1.

58 Everts, "Ecotoxicology for Risk Assessment in Arid Zones," 1.

59 Everts, "Ecotoxicology for Risk Assessment in Arid Zones," 1.

60 Rattner, "History of Wildlife Toxicology."

61 This history is not, of course, necessarily progressive: cuts to ecological and wildlife monitoring even in wealthy countries, notably the United States and Canada, have at times reversed the degree of environmental protection provided by governments.

62 For a similar argument, see Mika, "Fifty Years of Creativity."

5 · WAITING/NOT WAITING FOR POISON CONTROL

1 Alassane Diagne, personal communication, February 1, 2010.

2 Alassane Diagne, personal communication, February 15, 2010.

3 Fieldnotes, Dakar, February 1, 2010.

4 Mathias Camara, interview by author, February 15, 2010.

5 Fieldnotes, Dakar, February 15, 2010.

6 I attended weekly staff meetings; conversed, mostly informally, with staff members during their workday; participated in the design of and data collec-

tion for the envenomation survey; and hung out in the common office and, later, the helpline room.

7   Droney, "Ironies of Laboratory Work," 364.
8   Doudou Ba, interview by author, March 29, 2010.
9   Cailleux, "Identification rapide des formes médicamenteuses."
10 Kara, "Étude statistique des intoxications aigues"; Fall, "Pharmacovigilance."
11 "Compte-rendu de la seconde réunion du groupe de travail Centre Anti-Poison," November 22, 2001, LTH Administrative Archives.
12 UNITAR and IOMC, "UNITAR/IOMC Programme to Assist Countries."
13 WHO, Guidelines for Poison Control, v.
14 Govaerts et al., "Draft Report on the Survey of Poison Control Centres."
15 WHO, Guidelines for Poison Control, 3; Laborde, "New Roles for Poison Control Centres."
16 UNITAR and IOMC, "UNITAR/IOMC Programme to Assist Countries." The Rotterdam and Stockholm Conventions in particular create a demand for capacity, the first to undertake the evaluations that would allow for "Prior Informed Consent" to the importation of selected hazardous substances and the second to monitor the elimination of persistent organic pollutants (POPs). The Basel Convention is about controlling the circulation of hazardous waste.
17 UNITAR and IOMC, "UNITAR/IOMC Programme to Assist Countries." See also Clarke, "Experience of Starting a Poison Control Centre."
18 DEEC to Amadou Diouf, "Convocation à la réunion d'examen du projet d'arrêté portant création de la Commission Nationale de Gestion des Produits Chimiques," 2001, LTH Administrative Archives.
19 Amadou Diouf, "Plan d'action centre antipoison," n.d., photocopy of slide-show presentation, LTH Administrative Archives.
20 "DECRET n° 2004–1404 du 4 novembre 2004 portant organisation du Ministère de la Santé et de la Prévention Médicale," Journal Officiel de la République du Sénégal 6194 (December 4, 2004).
21 Accord de coopération scientifique entre l'UCAD (Sénégal) et l'Université du Littoral Côte d'Opale (France), 2004, LTH Administrative Archives.
22 Bierschenk, States at Work, 14.
23 This and the following quotes are from Alassane Diagne, interview by author, February 15, 2010.
24 Bierschenk, States at Work, 14.
25 Galvan, "Political Turnover"; Diouf, "Le clientélisme, la 'technocratie' et après?"
26 De Jong and Foucher, "La tragédie du roi Abdoulaye?"
27 De Jong and Quinn, "Ruines d'utopies."
28 De Jong and Foucher, "La tragédie du roi Abdoulaye?"; Melly, "Ethnography on the Road."
29 De Jong and Foucher, "La tragédie du roi Abdoulaye?," 192.
30 De Jong and Foucher, "La tragédie du roi Abdoulaye?," 193.

31 De Jong and Foucher, "La tragédie du roi Abdoulaye?," 188.
32 De Jong and Foucher, "La tragédie du roi Abdoulaye?," 190.
33 On migrants as investors in Dakar's future through partial acts of construction, see Melly, "Inside-out Houses."
34 République du Sénégal, "Plan national de développement sanitaire," 17.
35 Rees, "Humanity/Plan"; Redfield, *Life in Crisis*; Redfield, "Bioexpectations"; Nguyen, "Government-by-Exception." For further discussion of the historical and territorial distinctions between—and the different publics imagined and served by—global health and international and national health, see, for example, Brown et al., "The World Health Organization"; Lakoff, "Two Regimes of Global Health"; Prince, "Situating Health and the Public"; and Geissler, "Archipelago of Public Health."
36 Tousignant, "Citizens of Quality"; Tichenor, "Power of Data"; Léveque, "La gouvernementalié aux marges de l'Etat."
37 Geissler, *Para-States and Medical Science*.
38 Sullivan, "Mediating Abundance and Scarcity."
39 Geissler, "The Archipelago of Public Health."
40 Sullivan, "Mediating Abundance and Scarcity"; Tousignant, "Citizens of Quality"; Tichenor, "The Power of Data."
41 Mamdani, *Scholars in the Marketplace*, 98.
42 On the history and functioning of the LNCM, see Tousignant, "Pharmacy, Money and Public Health."
43 See also Tichenor, "Power of Data."
44 Note that Pure Earth (formerly the Blacksmith Institute), through the creation of a Global Alliance on Health and Pollution, has been actively campaigning for the recognition of pollution, including toxic exposure, as an urgent global health problem.

EPILOGUE

1 These have since been published as *Speculative Markets* and *Pharmacists in Senegal*. Additionally, Laurence Monnais and I jointly developed another project on pharmacists' expertise more broadly, in both Vietnam and Senegal; see "The Values of Versatility."
2 See the articles in Geissler and Tousignant, eds., *Capacity as History and Horizon*.
3 Bierschenk, *States at Work*, 2. On the privatization/criminalization and predatory nature of the state in Africa, see Bayart et al., *Criminalization of the State*; Mbembe, *On the Postcolony*; and Mbembe and Meintjes, "Necropolitics."
4 Bierschenk, *States at Work*, 9.

Abraham, Itty. "The Contradictory Spaces of Postcolonial Techno-science." *Economic and Political Weekly* 41, no. 3 (January 2006): 210–217.

Abraham, Itty. "Landscape and Postcolonial Science." *Contributions to Indian Sociology* 34, no. 2 (2000): 163–187.

Abraham, Itty. "Science and Power in the Postcolonial State." *Alternatives: Global, Local, Political* 21, no. 3 (1996): 321–339.

Adas, Michael. "Scientific Standards and Colonial Education in British India and French Senegal." In *Science, Medicine and Cultural Imperialism*, edited by Teresa A. Meade and Mark Walker, 4–35. London: Macmillan, 1991.

Affa'a, Félix-Marie, and Thérèse Des Lierres. *L'Afrique noire face à sa laborieuse appropriation de l'université: Les cas du Sénégal et du Cameroun.* Québec: Presses de l'Université Laval, 2002.

Amin, Ash. "Surviving the Turbulent Future." *Environment and Planning D: Society and Space* 31, no. 1 (2013): 140–156.

Amin, Ash. "Telescopic Urbanism and the Poor." *City* 17, no. 4 (2013): 476–492.

Amnesty International and Greenpeace Netherlands. *The Toxic Truth about a Company Called Trafigura, a Ship Called the Probo Kola, and the Dumping of Toxic Waste in Côte d'Ivoire.* September 25, 2012. http://www.greenpeace.org /international/en/publications/Campaign-reports/Toxics-reports/The-Toxic -Truth/.

Anderson, Kim A., Dogo Seck, Kevin A. Hobbie, Anna Ndiaye Traore, Melissa A. McCartney, Adama Ndaye, Norman D. Forsberg, Theodore A. Haigh, and Gregory J. Sower. "Passive Sampling Devices Enable Capacity Building and Characterization of Bioavailable Pesticide along the Niger, Senegal and Bani Rivers of Africa." *Philosophical Transactions of the Royal Society B: Biological Sciences* 369, no. 1639 (2014): 20130110. doi: 10.1098/rstb.2013.0110.

Anderson, Warwick. "The Frozen Archive, or Defrosting Derrida." *Journal of Cultural Economy* 8, no. 3 (2015): 379–387.

Arnaut, Karel, and Jan Blommaert. "Chthonic Science: Georges Niangoran-Bouah and the Anthropology of Belonging in Côte d'Ivoire." *American Ethnologist* 36, no. 3 (2009): 574–590.

Arnold, David. "Nehruvian Science and Postcolonial India." *Isis* 104, no. 2 (2013): 360–370.

Arnold, David. *Toxic Histories: Poison and Pollution in Modern India*. Cambridge: Cambridge University Press, 2016.

Asante, Kwadwo Ansong, Tetsuro Agusa, Charles Augustus Biney, William Atuobi Agyekum, Mohammed Bello, Masanari Otsuka, Takaaki Itai, Shin Takahashi, and Shinsuke Tanabe. "Multi-trace Element Levels and Arsenic Speciation in Urine of e-Waste Recycling Workers from Agbogbloshie, Accra in Ghana." *Science of the Total Environment* 424 (2012): 63–73.

Attisso, Michel A. "Contrôle de la qualité des médicaments au Sénégal. Rapport de mission, 1er-22 juillet 1969." Organisation Mondiale de la Santé, Rapport AFRO 0225, AFR/PHARM/3, April 29, 1970. National Library of Senegal.

Ba, Amadou, Robert Schilling, Ousmane Ndoye, Mamadou Ndiaye, and Amadou Kane. "L'arachide." In *Bilan de la recherche agricole et agroalimentaire au Sénégal 1964–2004*, 163–188. Dakar: IRA, 2005.

Ba, Doudou, Mounirou Ciss, Jean-François Cooper, and Babacar Niane. "Recherche et dosage des traces de métaux lourds dans des organismes marins au Sénégal." In *First Workshop of Participants in the Joint FAO-IOC-WHO-IAEA-UNEP Project on Monitoring of Pollution in the Marine Environment of the West and Central African Region (WACAF/2–Pilot Phase)*, Annex 5.3. Intergovernmental Oceanographic Commission, Workshop report no. 41, Dakar, Senegal, October 28–November 1, 1985.

Banza, Célestin Lubaba Nkulu, Tim S. Nawrot, Vincent Haufroid, Sophie Decrée, Thierry De Putter, Erik Smolders, Benjamin Ilunga Kabyla, Oscar Numbi Luboya, Augustin Ndala Ilunga, Alain Mwanza Mutombo, and Benoit Nemery. "High Human Exposure to Cobalt and Other Metals in Katanga, a Mining Area of the Democratic Republic of Congo." *Environmental Research* 109, no. 6 (2009): 745–752.

Barthélémy, Pascale. "La professionalisation des africaines en AOF (1920–1960)." *Vingtième Siècle: Revue d'histoire* 75 (2002): 35–46.

Basel Action Network. "Whistle Blower's Corner/Lawrence Summer's 1991 World Bank Memo." Accessed August 12, 2015. http://ban.org/whistle/summers.html.

Bathily, Aissatou. "Aflatoxines dans les aliments: Recherche et dosage dans les huiltes de pression artisanale et leurs résidus d'extraction, essais de detoxification." Pharmacy thesis, Cheikh Anta Diop University, Dakar, 1998.

Bauman, Zigmund. *Wasted Lives: Modernity and Its Outcasts*. Cambridge: Polity Press, 2004.

Bayart, Jean-François, Stephen Ellis, and Béatrice Hibou. *The Criminalization of the State in Africa*. Bloomington: Indiana University Press, 1999.

Beck, Ulrich. *Risk Society: Towards a New Modernity*. Translated by Mark Ritter. London: Sage Publications, 1992.

Bempah, Crentsil Kofi, and Augustine Kwame Donkor. "Pesticide Residues in Fruits at the Market Level in Accra Metropolis, Ghana, a Preliminary Study." *Environmental Monitoring and Assessment* 175 (2011): 551–561.

Berman, Alex. "J. B. A. Chevallier, Pharmacist-Chemist, a Major Figure in Nineteenth-Century Public Health." *Bulletin of the History of Medicine* 52 (1978): 200–213.

Bernabé, Jean. "Senghor: Raison hellène, émotion nègre." In *Archipélies n. 2 Seng-horiana: éloge à l'un des pères de la négritude*, edited by Pierre Dumont, Corinne Mencé-Caster, and Raphaël Confiant, 115–125. Paris: Éditions Publibook, 2001.

Beye, Alioune, Pape Charles Sow, and Harold van der Valk. "Effets du fenitrothion sur les coleopteres epiges de l'agroecosysteme du mil au Sénégal." In *Projet Locustox: Effet sur l'environnement de la lutte antiacridienne, Vol. III*, edited by James W. Everts, Djibril Mbaye, Oumar Barry, and Wim C. Mullié, 143–158. Rome: FAO, 1998. http://cereslocustox.sn.

Bierschenk, Thomas. *States at Work in West Africa: Sedimentation, Fragmentation and Normative Double-Binds*. Working Paper no. 113. Mainz: Institut für Ethnologie und Afrikastudien, Johannes Gutenberg-Universität, 2010. Accessed September 15, 2016. http://www.ifeas.uni-mainz.de/Dateien/AP113.pdf.

Blacksmith Institute. "Project Completion Report: Used Lead Acid Battery Contamination—Dakar, Senegal." Accessed August 12, 2015. http://www.blacksmith institute.org/files/FileUpload/files/Project%20Completion%20Reports/Project %20Completion%20Report%20Senegal%20updated%20July%202013%20.pdf.

Blacksmith Institute and Global Alliance on Health and Pollution. "The Poisoned Poor: Toxic Chemicals Exposures in Low- and Middle-Income Countries." Report, updated September 2013. Accessed October 20, 2017. http://gahp.net /wp-content/uploads/2017/02/GAHPPoisonedPoor_Report-Sept-2013.pdf.

Blanc, Floriane, and Philippe Jaussaud. "Les leçons inaugurales de chimie des pharmaciens français." *Revue d'histoire de la pharmacie* 94 (2007): 41–56.

Blum, Françoise. "Sénégal 1968: Révolte étudiante et grève générale." *Revue d'histoire moderne et contemporaine* 59, no. 2 (2012): 144–177.

Bonneuil, Christophe, and Patrick Petitjean. "Les chemins de la création de l'ORSTOM, du Front Populaire à la Libération en passant par Vichy: Recherche scientifique et politique coloniale, 1936–1945." In *Les sciences hors d'occident au XXe siècle, Vol 2, Les sciences coloniales: Figures et institutions*, edited by Patrick Petitjean, 114–161. Paris: Éditions ORSTOM, 1996.

Boudia, Soraya. "Global Regulation: Controlling and Accepting Radioactivity Risks." *History and Technology* 23, no. 4 (2007): 389–406.

Boudia, Soraya, and Nathalie Jas, eds. *Powerless Science? Science and Politics in a Toxic World*. Oxford: Berghahn Books, 2014.

Boudia, Soraya and Nathalie Jas. "Risk and Risk Society in Historical Perspective." *History and Technology* 23, no. 4 (2007): 317–331.

Boudia, Soraya, and Nathalie Jas, eds. *Toxicants, Health and Regulation since 1945*. London: Pickering and Chatto, 2015.

Boudia, Soraya, and Sébastien Soubiran. "Scientists and Their Cultural Heritage: Knowledge, Politics and Ambivalent Relationships." *Studies in History and Philosophy of Science Part A* 44, no. 4 (2013): 643–651.

Broadhead, Robert S. "Officer Ugg, Mr Yuk, Uncle Barf . . . Ad Nausea: Controlling Poison Control, 1950–1985." *Social Problems* 33, no. 5 (1986): 424–437.

Bromley, Simon. "Making Sense of Structural Adjustment." *Review of African Political Economy* 22 (1995): 339–348.

Brooke, James. "Waste Dumpers Turning to West Africa." *New York Times*, July 17, 1988. Accessed October 20, 2017. http://www.nytimes.com/1988/07/17/world /waste-dumpers-turning-to-west-africa.html?pagewanted=all.

Brown, Hannah. "Global Health Partnerships, Governance, and Sovereign Responsibility in Western Kenya." *American Ethnologist* 42, no. 2 (2015): 340–355.

Brown, Theodore M., Marcos Cueto, and Elizabeth Fee. "The World Health Organization and the Transition from "International" to "Global" Public Health." *American Journal of Public Health* 96, no. 1 (2006): 62–72.

Bull, David. *A Growing Problem: Pesticides and the Third World Poor*. Oxford: Oxfam, 1982.

Bullard, Robert D. *Dumping in Dixie: Race, Class, and Environmental Quality*. 3rd ed. Boulder, CO: Westview Press, [1990] 2000.

Cabral, Mathilde, Denis Dieme, Aminata Touré, Cheikh Diop, Fatmé Jichi, Fabrice Cazier, Mamadou Fall, and Amadou Diouf. "Impact du recyclage des batteries de véhicules sur la santé humaine et l'environnement: Étude pilote effectuée sur des femmes de Colobane et des mécaniciens de Médina." *Annales de toxicologie analytique* 24 (2012): 1–7.

Cabral, Mathilde, Denis Dieme, Anthony Verdin, Guillaume Garçon, Mamadou Fall, Saâd Bouhsina, Dorothée Dewaele, Fabrice Cazier, Aminata Tall-Dia, Amadou Diouf, and Pirouz Shirali. "Low-Level Environmental Exposure to Lead and Renal Adverse Effects: A Cross-Sectional Study in the Population of Children Bordering the Mbeubeuss Landfill near Dakar, Senegal." *Human and Experimental Toxicology* 31, no. 12 (2012): 1280–1292.

Cailleux, Michel. "Identification rapide des formes médicamenteuses solides (comprimés, gélules, capsules, cachets, pilules, etc.) autorisées au Sénégal: Contribution à la réalisation d'un fichier et d'une échantillonthèque pour le futur Centre Anti-Poisons de Dakar, en vue du traitement des intoxications médicamenteuses aigues." Pharmacy thesis, Université de Paris IX, 1984.

Caravanos, Jack, Edith E. Clarke, Carl S. Osei, and Yaw Amoyaw-Osei. "Exploratory Health Assessment of Chemical Exposures at E-Waste Recycling and Scrapyard Facility in Ghana." *Journal of Health and Pollution* 3 no. 4 (2013): 11–22.

Ceccaldi, Pierre Fernand. "Evolution de la toxicologie contemporaine." *Médecine d'Afrique Noire* 2, no. 48 (1956): 3–5.

Césaire, Georges, Robert Monnet, and François Fauran. "Les intoxications au Sénégal." *Médecine d'Afrique Noire* 18, no. 12 (1971): 895–914.

Chabas, Jean. "L'Institut des Hautes Etudes de Dakar." *Civilisations* 4, no. 2 (1954): 263–265.

Chindah, Alex Chuks, Amabaraye Solomon Braide, and Onyebuchi Cordelia Sibeudu. "Distribution of Hydrocarbons and Heavy Metals in Sediment and a Crustacean (Shrimps-Penaeus notialis) from the Bonny/New Calabar River Estuary, Niger Delta." *African Journal of Environmental Assessment and Management* 9 (2004): 1–17.

Clapp, Jennifer. *Toxic Exports: The Transfer of Hazardous Wastes from Rich to Poor Countries*. Ithaca, NY: Cornell University Press, 2001.

Clarke, Edith E. K. "The Experience of Starting a Poison Control Centre in Africa—The Ghana Experience." *Toxicology* 198, no. 1 (2004): 267–272.

Clarke, Sabine. "A Technocratic Imperial State? The Colonial Office and Scientific Research, 1940–1960." *Twentieth Century British History* 18, no. 4 (2007): 453–480.

Collin, Johanne. *Changement d'ordonnance: Mutations professionnelles, identité sociale et féminisation de la profession pharmaceutique au Québec, 1940–1980.* Montreal: Éditions Boréal, 1995.

Collomb, Henri, Jacques Zwingelstein, and M. Picca. "Les conduites suicidaires à Dakar en milieu non africain." *Bulletins et mémoires de la Faculté Mixte de Médecine et de Pharmacie de Dakar* 10 (1962): 156–161.

Coly, Alpha. "Recherche et dosage des résidus de pesticides organochlorés dans le lait maternel au niveau de la région de Dakar." Pharmacy thesis, Université Cheikh Anta Diop, Dakar, 2000.

Cooper, Frederick. "Africa's Pasts and Africa's Historians." *Canadian Journal of African Studies/La revue canadienne des études africaines* 34, no. 2 (2000): 297–336.

Cooper, Frederick. "Conflict and Connection: Rethinking Colonial African History." *American Historical Review* 99, no. 5 (1994): 1516–1545.

Cooper, Frederick. *Decolonization and African Society: The Labor Question in French and British Africa.* Cambridge: Cambridge University Press, 1996.

Crane, Johanna T. *Scrambling for Africa: AIDS, Expertise, and the Rise of American Global Health Science.* Ithaca, NY: Cornell University Press, 2013.

Crane, Johanna T. "Unequal 'Partners': AIDS, Academia, and the Rise of Global Health." *Behemoth: A Journal on Civilisation* 3, no. 3 (2010): 78–97.

Cruise O'Brien, Rita. *White Society in Black Africa: The French of Senegal.* London: Faber and Faber, 1972.

Daemmrich, Arthur. "Forum: Risk Frameworks and Biomonitoring: Distributed Regulation of Synthetic Chemicals in Humans." *Environmental History* 13, no. 4 (2008): 684–693.

Dalberto, Séverine Awenengo, Hélène Charton, and Odile Goerg. "Urban Planning, Housing and the Making of 'Responsible Citizens' in the Late Colonial Period: Dakar, Nairobi and Conakry." In *Governing Cities in Africa*, edited by Simon Bekker and Laurent Fourchard, 43–66. Cape Town: HSRC Press, 2013.

Danfa, Abdoulaye, Baba Fall, and Harold van der Valk. "Tests de toxicité aigue sur un parasitoide, bracon hebetor say (hymenoptera: braconidae), avec différents insecticides utilisés en lutte antiacridienne au Sahel." In *Projet Locustox: Effet sur l'environnement de la lutte antiacridienne, Vol. II*, edited by James W. Everts, Djibril Mbaye, Oumar Barry, and Wim C. Mullié, 117–136. Rome: FAO, 1998.

Danfa, Abdoulaye, and Harold van der Valk. "Toxicity Tests with Fenitrothion on Pimelia Senegalensis and Trachyderma Hispida (Coleoptera Tenedrionidae)." In *Projet Locustox: Effet sur l'environnement de la lutte antiacridienne, Vol. I*, edited by James W. Everts, Djibril Mbaye, and Oumar Barry, 161–173. Rome: FAO, 1997.

De Boeck, Filip. "'Divining' the City: Rhythm, Amalgamation and Knotting as Forms of 'Urbanity.'" *Social Dynamics: A Journal of African Studies* 41, no. 1 (2015): 47–58.

De Boeck, Filip. "Infrastructure: Commentary from Filip De Boeck." Curated Collections, *Cultural Anthropology Online*, November 26, 2012. http://www .culanth.org/curated_collections/11-infrastructure/discussions/7-infrastructure -commentary-from-filip-de-boeck.

De Boeck, Filip. "Postcolonialism, Power and Identity: Local and Global Perspectives from Zaire." In *Postcolonial Identities in Africa*, edited by Richard P. Werbner and Terence O. Ranger, 75–106. London: Zed Books, 1996.

Debue-Barazer, Christine. "Des simples aux plantes médicinales: Emile Perrot (1867–1951), un pharmagnoste colonial." Advanced Studies Diploma, Université Paris IV-Sorbonne, 2002.

De Jong, Ferdinand, and Brian Quinn. "Ruines d'utopies: L'École William Ponty et l'Université du Futur Africain." *Politique africaine* 135 (2014): 71–94.

De Jong, Ferdinand, and Michael Rowlands, eds. *Reclaiming Heritage: Alternative Imaginaries of Memory in West Africa.* Walnut Creek, CA: Left Coast Press, 2009.

De Jong, Ferdinand, and Vincent Foucher. "La tragédie du roi Abdoulaye? Néo-modernisme et Renaissance africaine dans le Sénégal contemporain." *Politique africaine* 2 (2010): 187–204.

De Jorio, Rosa. "Introduction to Special Issue: Memory and the Formation of Political Identities in West Africa." *Africa Today* 52, no. 4 (2006): v–ix.

Derrida, Jacques. "Archive Fever: A Freudian Impression." *Diacritics* 25, no. 2 (1995): 9–63.

DeSilvey, Caitlin, and Tim Edensor. "Reckoning with Ruins." *Progress in Human Geography* 37, no. 4 (2013): 465–485.

Dia, Yaya Mamadou. "Recherche et dosage des aflatoxines dans l'huile d'arachide de pression préparée artisanalement dans les régions de Diourbel et Kaolack (Sénégal)." Pharmacy thesis, Université Cheikh Anta Diop, Dakar, 1997.

Diarra, Amadou. "Pesticides Registration and Regulation in CILSS Member Countries." *Pesticides Management in West Africa* 8 (2011): 25–29.

Diatta, Thierno. "Contribution à l'étude de la qualité des corps gras alimentaires commercialisés au Sénégal: Les huiles végétales." Pharmacy thesis, Université Cheikh Anta Diop, Dakar, 1998.

Dieme, Denis, Mathilde Cabral, Anthony Verdin, Mamadou Fall, Sylvain Billet, Fabrice Cazier, Guillaume Garçon, Amadou Diouf, and Pirouz Shirali. "Caractérisation physico-chimique et effets cytotoxiques de particules atmosphériques $PM2,5$ de la ville de Dakar (Sénégal)." *Annales de Toxicologie Analytique* 23, no. 4 (2011): 157–167.

Diop, Amadou. "Etude de la qualité de l'eau dans le district rural de Khombole." Pharmacy thesis, Université Cheikh Anta Diop, Dakar, 1995.

Diop, Amadou, Yérim M. Diop, Diène D. Thiaré, Fabrice Cazier, Serigne O. Sarr, Amaury Kasprowiak, David Landy, and François Delattre. "Monitoring Survey of the Use Patterns and Pesticide Residues on Vegetables in the Niayes Zone, Senegal." *Chemosphere* 144 (2016): 1715–1721.

Diop, Cheikh, Dorothée Dewaele, Aminata Toure, Mathilde Cabral, Fabrice Cazier, Mamadou Fall, Baghdad Ouddane, and Amadou Diouf. "Étude de la contamination par les éléments traces métalliques des sédiments côtiers au niveau des points d'évacuation des eaux usées à Dakar (Sénégal)." *Revue des sciences de l'eau/Journal of Water Science* 25, no. 3 (2012): 277–285.

Diop, Cheikh Anta. "Apport de l'Afrique à la Civilisation Universelle dans le domaine des sciences exactes." In *Sciences et philosophie: Textes 1960–1986*, edited by Cheikh Mbacké Diop, 53–72. Dakar: IFAN, 2007.

Diop, Cheikh Anta. "Perspectives de la recherche scientifique en Afrique." In *Sciences et philosophie: Textes 1960–1986*, edited by Cheikh Mbacké Diop, 123–131. Dakar: IFAN, 2007.

Diop, Momar Coumba. "Essai sur 'l'art de gouverner' le Sénégal." In *Gouverner le Sénégal: Entre ajustement structurel et développement durable*, edited by Momar Coumba Diop, 9–39. Dakar: Éditions Karthala, 2004.

Diop, Yérim M., Amadou Diouf, Mamadou Fall, A. Thiam, Bara Ndiaye, Mounirou Ciss, and Doudou Ba. "Bioaccumulation des pesticides: Recherche et dosage de résidus d'organochlorés dans les produits d'origine végétale." *Dakar médical* 44, no. 2 (1998): 153–157.

Diop, Yérim M., Bara Ndiaye, Amadou Diouf, Mamadou Fall, C. Thiaw, A. Thiam, Oumy Barry, Mounirou Ciss, and Doudou Ba. "Contamination par les aflatoxines des huiles d'arachide artisanales préparées au Sénégal." *Annales pharmaceutiques françaises* 58 (2000): 470–474.

Diouf, Amadou, Guillaume Garçon, C. Thiaw, Yérim M. Diop, Mamadou Fall, Bara Ndiaye, T. Siby, et al. "Environmental Lead Exposure and Its Relationship to Traffic Density among Senegalese Children: A Pilot Study." *Human and Experimental Toxicology* 22, no. 10 (2003): 559–564.

Diouf, Amadou, Guillaume Garçon, Yerim M. Diop, Bara Ndiaye, C. Thiaw, Mamadou Fall, Oumy Kane-Barry, Doudou Ba, J. M. Haguenoer, and Pirouz Shirali. "Environmental Lead Exposure and Its Relationship to Traffic Density among Senegalese Children: A Cross-Sectional Study." *Human and Experimental Toxicology* 25, no. 11 (2006): 637–644.

Diouf, Amadou, Mounirou Ciss, Yerim Diop, C. S. Boye, S. Diouf, Mamadou Fall, Amadou Diop, and Doudou Ba. "Etude du niveau de pollution de l'eau de puits du district de Khombole: Recherche de contamination par les résidus de pesticides organochlorés et par les matières organiques (fécés)." *Dakar medical* 43, no. 2 (1997): 157–160.

Diouf, Amadou, Yérim M. Diop, Bara Ndiaye, Mamadou Fall, D. Sarr, A. Thiam, Oumy Barry, C. Thiaw, Doudou Ba, and Mounirou Ciss. "Utilisation des feuilles de manguier (Manguifera indica, Anacardiacea) comme bioindicateur de la pollution atmosphérique par le pp'Dichlorodiphenyltrichloroethane (pp'DDT)." *Dakar médical* 45, no. 2 (1999): 122–125.

Diouf, E., B. Sali Ka, O. Kane, P. N. Dieng, and J. F. Dienne. "Intoxication massive à l'ammoniac dans un pays en voie de développement: Bilan lésionnel et facteurs de gravité." *Réanimation urgences* 8, no. 8 (1999): 633–637.

Diouf, Mamadou. "Le clientélisme, la 'technocratie' et après?" In *Sénégal: Trajectoires d'un État*, edited by Momar Coumba Diop, 233–278. Dakar: Codesria, 1992.

Diouf, Mamadou. "Fresques murales et écriture de l'histoire. Le Set/Setal à Dakar." *Politique africaine* 46 (1992): 41–54.

Diouf, Mamadou, ed. *L'historiographie indienne en débat: colonialisme, nationalisme et sociétés postcoloniales.* Paris: Éditions Karthala, 1999.

Diouf, Mamadou. "Senegalese Development: From Mass Mobilization to Technocratic Elitism." In *International Development and the Social Sciences: Essays on the History and Politics of Knowledge,* edited by Frederick Cooper and Randall M. Packard, 291–319. Berkeley: University of California Press, 1997.

Diouf, Mamadou. "Urban Youth and Senegalese Politics: Dakar 1988–1994." *Public Culture* 8 (1996): 225–250.

Diouf, Mamadou Lamine. "Aflatoxines dans l'arachide: Essai de décontamination de l'huile d'arachide de pression artisanale par exposition aux rayons solaires." Pharmacy thesis, Université Cheikh Anta Diop, Dakar, 2001.

Droney, Damien. "Ironies of Laboratory Work during Ghana's Second Age of Optimism." *Cultural Anthropology* 29, no. 2 (2014): 363–384.

Dzidzornu, David M. "Marine Pollution Control in the West and Central African Region." *Queen's Law Journal* 20 (1994): 439–486.

Echenberg, Myron J. *Black Death, White Medicine: Bubonic Plague and the Politics of Public Health in Colonial Senegal, 1914–1945.* Portsmouth, NH: Heinemann, 2002.

Edensor, Tim. "The Ghosts of Industrial Ruins: Ordering and Disordering Memory in Excessive Space." *Environment and Planning D: Society and Space* 23, no. 6 (2005): 829–849.

Edwards, C. Richard, and Ellis W. Huddleston. *Efficacy and Environmental Effects of Large Plane and Small Plane Operations in Senegal and Proposed Plan for Gathering Information for 1987 Environmental Assessment.* Washington, DC: USAID, 1986.

Eisemon, Thomas Owen, and Charles H. Davis. "Can the Quality of Scientific Training and Research in Africa Be Improved?" *Minerva* 29 (1991): 1–26.

Eisemon, Thomas Owen, and Jamil Salmi. "African Universities and the State: Prospects for Reform in Senegal and Uganda." *Higher Education* 25, no. 2 (1993): 151–168.

Elderkin, Susan, Richard Wiles, and Christopher Campbell. *Forbidden Fruit: Illegal Pesticides in the US Food Supply.* Washington, DC: Environmental Working Group, 1995.

Emoyan, Onoriode O., F. E. Ogban, and E. Akarah. "Evaluation of Heavy Metals Loading of River Ijana in Ekpan-Warri, Nigeria." *Journal of Applied Sciences and Environmental Management* 10, no. 2 (2006): 121–127.

Environmental Defense Fund. *Toxic Ignorance: The Continuing Absence of Basic Health Testing for Top-selling Chemicals in the United States.* New York: Environmental Defense Fund, 1997.

Everts, James W. "Ecotoxicology for Risk Assessment in Arid Zones: Some Key Issues." *Archives of Environmental Contamination and Toxicology* 32, no. 1 (1997): 1–10.

Everts, James W. *Environmental Effects of Chemical Locust and Grasshopper Control: A Pilot Study.* Rome: FAO, 1990.

Everts, James W., Dibril Mbaye and Oumar Barry, eds. *Locustox Project: Environmental Side-Effects of Locust and Grasshopper Control, Vol. I.* Rome: FAO, 1997.

Fall, Abdoulaye Dieugue. "Pharmacovigilance: Aperçu général et problèmes spécifiques au Sénégal." Pharmacy thesis, Université Cheikh Anta Diop, Dakar, 1985.

Fall, Abdou Salam, Babacar Gueye, Djiby Diakhate, Omar Saïp Sy, Abdoulaye Dieye, Sémou Ndiaye, Yaya Bodian, and Abdoulaye Sakho. "Gouvernance et corruption dans le système de santé au Sénégal: Rapport provisoire." Dakar: Forum Civil/CRDI, 2004.

Fassin, Didier. "Du clandestin à l'officieux: Les réseaux de vente illicite des médicaments au Sénégal." *Cahiers d'études africaines* (1985): 161–177.

Fassin, Didier. "La vente illicite des médicaments au Sénégal: Économies 'parallèles,' état et société." *Politique africaine* 23 (1986): 123–130.

Fauran, François, Catherine Pellissier, and Babacar Niane. "Dosage de l'Alpha chloralose dans un raticide commercial." *Afrique médicale* 10, no. 92 (1971): 647.

Fauran, François, Catherine Pellissier, and S. Fabre. "Métabolisme du chloralose Alpha I: Dosage colorimétrique du chloralose libre et glucuroconjugué dans les urines." *Annales pharmaceutiques françaises* 30, no. 4 (1972): 289–298.

Fauran, François, Catherine Pellissier, S. Fabre, and Babacar Niane. "Métabolisme du chloralose Alpha II Etude cinétique de l'élimination urinaire des formes libres et glucuroconjuguées chez le rat." *Annales pharmaceutiques françaises* 30, no. 5 (1972): 373–378.

Fauran, François, Catherine Pellissier, S. Fabre, and Georges Gras. "Dosage colorimétrique du chloralose Alpha dans les urines." *Produits et problèmes pharmaceutiques* 25, no. 11 (1970): 881–886.

Fauran, François, Georges Césaire, Catherine Pellissier, and J. Goudote. "A propos d'une intoxication mortelle par la chloroquine: Intérêt analytique de la détection des metabolites." *Annales pharmaceutiques françaises* 28, no. 11 (1970): 667–674.

Fauran, François, and Georges Gras. "Etude spectrophotometrique des réactions de complexation des diacetates de dialkylplomb par la diphenylthiocarbazone (dithizione)." *Analusis* 4, no. 1 (1976): 34–40.

Ferguson, James. "Decomposing Modernity: History and Hierarchy after Development." In *Postcolonial Studies and Beyond*, edited by Ania Loomba, Suvir Kaul, and Matti Bunzl, 166–181. Durham, NC: Duke University Press, 2005.

Ferguson, James. *Expectations of Modernity: Myths and Meanings of Urban Life on the Zambian Copperbelt.* Berkeley: University of California Press, 1999.

Ferguson, James. *Global Shadows: Africa in the Neoliberal World Order.* Durham, NC: Duke University Press, 2006.

Foley, Ellen E. *Your Pocket Is What Cures You: The Politics of Health in Senegal.* New Brunswick, NJ: Rutgers University Press, 2009.

Food and Agricultural Organization (FAO). *The Desert Locust Research and Development Register*, no. 2, March 1990.

Food and Agricultural Organization (FAO). *Effets sur l'environnement de la lutte antiacridienne. Senegal: Conclusions et recommandations du projet.* Rome: FAO, 1998.

Food and Agricultural Organization (FAO). *Rapport de la vingt-neuvième session du Comite FAO de lutte contre le criquet pelerin*. Rome: FAO, 1988.

Food and Agricultural Organization (FAO). "Thirty-Second Session, Rome, 29 November–10 December 2003, Presentation of the B. R. Sen Awards." C 2003 /INF/6, November 2003, FAO Corporate Document Repository. http://www.fao .org/docrep/meeting/007/J0779e.htm.

Fortun, Kim. *Advocacy after Bhopal: Environmentalism, Disaster, New Global Orders*. Chicago: University of Chicago Press, 2009.

Fortun, Kim, and Mike Fortun. "Scientific Imaginaries and Ethical Plateaus in Contemporary US Toxicology." *American Anthropologist* 107, no. 1 (2005): 43–54.

Freidberg, Susanne. *French Beans and Food Scares: Culture and Commerce in an Anxious Age*. Oxford: Oxford University Press, 2004.

Frickel, Scott. *Chemical Consequences: Environmental Mutagens, Scientist Activism, and the Rise of Genetic Toxicology*. New Brunswick, NJ: Rutgers University Press, 2004.

Frickel, Scott, and M. Bess Vincent. "Hurricane Katrina, Contamination, and the Unintended Organization of Ignorance." *Technology in Society* 29, no. 2 (2007): 181–188.

Fullwiley, Duana. *The Encultured Gene: Sickle Cell Health Politics and Biological Difference in West Africa*. Princeton, NJ: Princeton University Press, 2011.

Gaillard, Jacques. "The Senegalese Scientific Community: Africanization, Dependence and Crisis." In *Scientific Communities in the Developing World*, edited by Jacques Gaillard, V. V. Krishna, and Roland Waast, 155–182. New Delhi: Sage Publications, 1997.

Gaillard, Jacques, Mohamed Hassan, and Roland Waast, with Daniel Schaffer. "Africa." In *UNESCO Science Report 2005*, edited by Mustafa El Tayeb and S. Schneegans, 177–201. Paris: UNESCO Publishing, 2005.

Gaillard, Jacques, and Roland Waast. "L'aide à la recherche en Afrique subsaharienne: Comment sortir de la dépendance? Le cas du Sénégal et de la Tanzanie." *Autrepart* 13 (2000): 71–89.

Gaillard, Jacques, and Roland Waast. "La recherche scientifique en Afrique." *Afrique contemporaine* 148 (1988): 3–30.

Galt, Ryan E. "Beyond the Circle of Poison: Significant Shifts in the Global Pesticide Complex, 1976–2008." *Global Environmental Change* 18 (2008): 786–799.

Galvan, Dennis Charles. "Political Turnover and Social Change in Senegal." *Journal of Democracy* 12, no. 3 (2001): 51–62.

Gaye, Makhoudia. "Contribution à l'évaluation de la performance des étudiants en pharmacie aux travaux pratiques." Pharmacy thesis, Université Cheikh Anta Diop, Dakar, 1995.

Geissler, P. Wenzel. "The Archipelago of Public Health: Comments on the Landscape of Medical Research in Twenty-First-Century-Africa." In *Making and Unmaking Public Health in Africa*, edited by Ruth Prince and Rebecca Marsland, 231–256. Athens: Ohio University Press, 2014.

Geissler, P. Wenzel. "Parasite Lost: Remembering Modern Times with Kenyan

Government Medical Scientists." In *Evidence, Ethos and Experiment: The Anthropology and History of Medical Research in Africa*, edited by P. Wenzel Geissler and Catherine Molyneux, 297–332. New York: Berghahn Books, 2011.

Geissler, P. Wenzel, ed. *Para-States and Medical Science: Making African Global Health*. Durham, NC: Duke University Press, 2015.

Geissler, P. Wenzel. "What Future Remains? Remembering an African Place of Science." In *Para States and Medical Science: Making African Global Health*, edited by P. Wenzel Geissler, 142–178. Durham, NC: Duke University Press, 2015.

Geissler, P. Wenzel, Ann H. Kelly, John Manton, Ruth J. Prince, and Noémi Tousignant. "Introduction: Sustaining the Life of the Polis." *Africa* 83 (2013): 531–538.

Geissler, P. Wenzel, Guillaume Lachenal, John Manton, and Noémi Tousignant, eds. *Traces of the Future: An Archaeology of Medical Science in Twenty-First Century Africa*. Bristol: Intellect Press, 2016.

Geissler, P. Wenzel, and Noémi Tousignant. "Capacity as History and Horizon: Infrastructure, Autonomy and Future in African Health Science and Care." *Canadian Journal of African Studies/Revue canadienne des études africaines* 50, no. 3 (2016): 349–359.

Geissler, P. Wenzel, and Noémi Tousignant, eds. Special Issue: Capacity as History and Horizon. *Canadian Journal of African Studies/Revue canadienne des études africaines* 50, no. 3 (2016): 349–478.

Gieryn, Thomas F. "Laboratory Design for Post-Fordist Science." *Isis* 99, no. 4 (2008): 796–802.

Gilbert, Hannah. "Re-visioning Local Biologies: HIV-2 and the Pattern of Differential Valuation in Biomedical Research." *Medical Anthropology* 32, no. 4 (2013): 343–358.

Global Pharma Health Fund E.V. "Minilab—Protection against Counterfeit Medicines." Accessed August 7, 2014. http://www.gphf.org/web/en/minilab.

González-Ruibal, Alfredo. "The Dream of Reason: An Archaeology of the Failures of Modernity in Ethiopia." *Journal of Social Archaeology* 6, no. 2 (2006): 175–201.

Gooday, Graeme. "Placing or Replacing the Laboratory in the History of Science?" *Isis* 99, no. 4 (2008): 783–795.

Goudiaby, Jean-Alain. "Le Sénégal dans son appropriation de la réforme LMD: Déclinaison locale d'une réforme 'globale.'" *Journal of Higher Education in Africa/Revue de l'enseignement supérieur en Afrique* 7 (2009): 79–93.

Govaerts, M., L. Roche, A. Berlin, J. Haines, and M. Th. van der Venne. "Draft Report on the Survey of Poison Control Centres and Related Toxicological Services." CEE, World Federation of Associations of Clinical Toxicology Centres and Poison Control Centres, and IPCS, 1986.

Graham, Stephen, and Nigel Thrift. "Out of Order Understanding Repair and Maintenance." *Theory, Culture and Society* 24 no. 3 (2007): 1–25.

Grant, Ian. "Appendix 15." In *Rapport de la réunion sur la recherche antiacridienne "Définition des priorités futures en recherche."* Rome: FAO, 1988.

Gras, Georges. "La lutte chimique contre les moustiques: Étude de quelques problèmes posés par l'utilisation des insecticides en Languedoc-Roussillon." PhD diss., Université de Montpellier, 1968.

Gras, Georges, Catherine Pellissier, and François Fauran. "Toxicologie analytique du chloralose Alpha: Application dans 3 cas d'intoxication aigue." *European Journal of Toxicology and Environmental Hygiene* 8, no. 5 (1975): 371–377.

Gras, Georges, and François Fauran. "Dosage colorimétrique du diacétate de plomb dibutyle dans les urines." *Annales pharmaceutiques françaises* 30, nos. 7–8 (1972): 545–554.

Gras, Georges, and François Fauran. "Etude de la réaction de complexation du diacétate de plomb dibutyle (DPD) par la dithizone." *Annales pharmaceutiques françaises* 30 (1972): 459–472.

Gras, Georges, H. Giono, M. Graber, and C. Quénum. "Toxicologie d'un nouvel anthelminthique: Le diacétate de plomb dibutyle (DPD)." *Médecine d'Afrique noire* 18, no. 12 (1971): 921–922.

Gras, Georges, and Janine Mondain. "Etude experimentale de la bioaccumulation du mercure mineral et du methylmercure chez les poissons rouges (Carassius Auratus L.)." *Comptes rendus de la Société de biologie* 174, no. 5 (1980): 929.

Gras, Georges, and Janine Mondain. "Influence de la consommation de poisson sur la teneur en mercure des cheveux et du sang chez deux groupes sociaux sénégalais différents." *Revue internationale d'oceanographie médicale* 59 (1980): 63–70.

Gras, Georges, and Janine Mondain. "Microdosage colorimétrique du mercure total par la Di-Beta-Naphtylthiocarbazone: Application au contrôle dans le poisson." *Annales pharmaceutiques françaises* 39, no. 6 (1981): 529–536.

Gras, Georges, and Janine Mondain. "Problème posé par l'utilisation des cosmétiques mercuriels au Sénégal." *Toxicological European Research* 3 (1981): 175–178.

Gras, Georges, and Janine Mondain. "Rapport methylmercure/mercure total dans differentes espèces de poissons pêchés sur les côtes de l'Afrique de l'Ouest." *Toxicological European Research* 4, no. 14 (1982): 191.

Gras, Georges, and Janine Mondain. "Teneur en mercure de quelques espèces de poissons pêchés sur les côtes du Senegal." *Revue internationale d'oceanographie médicale* 51–52 (1978): 83–88.

Guyer, Jane I. "Prophecy and the Near Future: Thoughts on Macroeconomic, Evangelical and Punctuated Time." *American Ethnologist* 34, no. 3 (2007): 409–421.

Haddar, Mohamed, ed. *Actes du Séminaire sur les conséquences sociales de l'ajustement structurel au Sénégal.* Vol. 1. Dakar: IDEP, 1992.

Haefliger, Pascal, Monique Mathieu-Nolf, Stephanie Lociciro, Cheikh Ndiaye, Malang Coly, Amadou Diouf, Absa Lam Faye, Aminata Sow, Joanna Tempowski, Jenny Pronczuk, Antonio Pedre Filipe Junior, Roberto Bertollini, and Maria Neira. "Mass Lead Intoxication from Informal Used Lead-Acid Battery Recycling in Dakar, Senegal." *Environmental Health Perspectives* 117, no. 10 (2009): 1535–1540.

Harney, Elizabeth. *In Senghor's Shadow: Art, Politics, and the Avant-Garde in Senegal, 1960–1995.* Durham, NC: Duke University Press, 2004.

Harris, Verne. "The Archival Sliver: Power, Memory, and Archives in South Africa." *Archival Science* 2, nos. 1–2 (2002): 63–86.

Harvey, Penny, and Hannah Knox. "The Enchantments of Infrastructure." *Mobilities* 7, no. 4 (2012): 534.

Hecht, Gabrielle. *Being Nuclear: Africans and the Global Uranium Trade.* Cambridge, MA: MIT Press, 2012.

Hecht, Gabrielle. "Rupture-Talk in the Nuclear Age: Conjugating Colonial Power in Africa." *Social Studies of Science* 32, nos. 5–6 (2002): 691–727.

Hoffman, Karen. "Unheeded Science: Taking Precaution out of Toxic Water Pollutants Policy." *Science, Technology and Human Values* 38, no. 6 (2013): 829–850.

Hokkanen, Markku. "Imperial Networks, Colonial Bioprospecting and Burroughs Wellcome and Co.: The Case of Strophanthus Kombe from Malawi (1859–1915)." *Social History of Medicine* 25, no. 3 (2012): 589–607.

Hough, Peter. "Poisons in the System: The Global Regulation of Hazardous Pesticides." *Global Environmental Politics* 3, no. 2 (2003): 11–24.

Hountondji, Paulin. "Scientific Dependence in Africa Today." *Research in African Literatures* 21, no. 3 (1990): 5–15.

Ikenaka, Yoshinori, Shouta M. M. Nakayama, Kaampwe Muzandu, Kennedy Choongo, Hiroki Teraoka, Naoharu Mizuno, and Mayumi Ishizuka. "Heavy Metal Contamination of Soil and Sediment in Zambia." *African Journal of Environmental Science and Technology* 4, no. 11 (2010): 729–739.

Ikingura, Justinian R., and Hirokatsu Akagi. "Monitoring of Fish and Human Exposure to Mercury Due to Gold Mining in the Lake Victoria Goldfields, Tanzania." *Science of the Total Environment* 191, no. 1 (1996): 59–68.

Iliffe, John. *East African Doctors: A History of the Modern Profession.* Cambridge: Cambridge University Press, 1998.

Irissou, Louis. "Brève histoire de la toxicologie: M. le Professeur Roger Douris, Leçon de réouverture du cours de la toxicologie à la Faculté de pharmacie de Nancy, 7 mars 1946." *Revue d'histoire de la pharmacie* 35 (1947): 163–165.

Isaacman, Allen F., Premesh Lalu, and Thomas I. Nygren. "Digitization, History, and the Making of a Postcolonial Archive of Southern African Liberation Struggles: The Aluka Project." *Africa Today* 52, no. 2 (2005): 55–77.

Jackson, Jeffrey J. "FAO Quelea Research in Africa." *Bird Control Seminars Proceedings,* Paper 126 (1973): 221–222. http://digitalcommons.unl.edu/icwdmbird control/126/.

Jaffré, Yannick, and Jean-Pierre Olivier de Sardan. *Une médecine inhospitalière: Les difficiles relations entre soignants et soignés dans cinq capitales d'Afrique de l'Ouest.* Paris: Éditions Karthala, 2003.

Jardine, Nicholas, and Lydia Wilson. "Recent Material Heritage of the Sciences." *Studies in History and Philosophy of Science Part A* 44, no. 4 (2013): 632–633.

Jas, Nathalie. "Public Health and Pesticide Regulation in France before and after Silent Spring." *History and Technology* 23, no. 4 (2007): 369–388.

Kandji, Ndeye Anta. "Etude de la composition chimique et de la qualité d'huiles végétales artisanales consommées au Sénégal." Pharmacy thesis, Université Cheikh Anta Diop, Dakar, 2001.

Kane, Mamadou Malick. "Contribution à l'état de la qualité et des fraudes du lait frais fermier commercialité sur le marché de Dakar." Pharmacy thesis, Université Cheikh Anta Diop, Dakar, 2001.

Kane, Serigne Mor. "Contribution à l'étude de la qualité des beurres vendus à Dakar (Senegal)." Pharmacy thesis, Université Cheikh Anta Diop, Dakar, 2001.

Kara, Soundiagaussé. "Étude statistique des intoxications aigües du point de vue hospitalier." Thesis of medicine, Université Cheikh Anta Diop, Dakar, 1984.

Kerharo, Joseph, and Armand Bouquet. *Plantes médicinales et toxiques de la Côte-d'Ivoire-Haute-Volta*. Paris: Vigot Frères, 1950.

Kilroy-Marac, Kathleen. "The Impossible Inheritance: Memory and Postcolonial Subjectivity at the Fann Psychiatric Clinic in Dakar, Senegal." PhD diss., Columbia University, 2010.

Kilroy-Marac, Kathleen. "Nostalgic for Modernity: Reflecting on the Early Years of the Fann Psychiatric Clinic in Dakar, Senegal." *African Identities* 11, no. 4 (2013): 367–380.

Kinyamu, J. K., Laetitia W. Kanja, J. U. Skaare, and T. E. Maitho. "Levels of Organochlorine Pesticide Residues in Milk of Urban Mothers in Kenya." *Bulletin of Environmental Contamination and Toxicology* 60, no. 5 (1998): 732–738.

Kohler, Robert E. "Lab History: Reflections." *Isis* 99, no. 4 (2008): 761–768.

Koné, Lassana. "Pollution in Africa: A New Toxic Waste Colonialism? An Assessment of Compliance of the Bamako Convention in Côte d'Ivoire." LLM diss., University of Pretoria, 2009.

Kowal, Emma, and Joanna Radin. "Indigenous Biospecimen Collections and the Cryopolitics of Frozen Life." *Journal of Sociology* 51, no. 1 (2015): 63–80.

Kusiak, Pauline. "Instrumentalized Rationality, Cross-Cultural Mediators, and Civil Epistemologies of Late Colonialism." *Social Studies of Science* 40, no. 6 (2010): 871–902.

Kusiak, Pauline. "'Tubab' Technologies and 'African' Ways of Knowing: Nationalist Techno-Politics in Senegal." *History and Technology* 26, no. 3 (2010): 227.

Laborde, Amalia. "New Roles for Poison Control Centres in the Developing Countries." *Toxicology* 198, no. 1 (2004): 273–277.

Lachenal, Guillaume. "The Intimate Rules of the French Coopération: Morality, Race and the Postcolonial Division of Scientific Work at the Pasteur Institute of Cameroon." In *Evidence, Ethos and Experiment: The Anthropology and History of Medical Research in Africa*, edited by P. Wenzel Geissler and Catherine Molyneux, 373–402. London: Berghahn Books, 2011.

Laffitte, Numa. *La pharmacopée indigène en Afrique occidentale française*. Edited by Armand Bouquet and Joseph Kerharo. Paris: Office de la recherche scientifique coloniale, 1946.

Lahr, Joost, Karim B. Ndour, Amadou Badji, and A. O. Diallo. *Effects of Experimental Locust Control with Deltamethrin and Bendiocarb on the Aquatic Invertebrate Fauna of Temporary Ponds in Central Senegal*. Locustox Report 95/3. FAO, Locustox Project, Dakar, Senegal, 1995.

Lakoff, Andrew. "Two Regimes of Global Health." *Humanity* 1, no. 1 (2010): 59–79.

Langston, Nancy. *Toxic Bodies: Hormone Disruptors and the Legacy of DES*. New Haven, CT: Yale University Press, 2010.

Lee, Raymond. "Weber, Re-enchantment and Social Futures." *Time and Society* 19, no. 2 (2010): 180–192.

Le Moan, Georges. "L'enseignement de la toxicologie à Paris dans le cursus des études pharmaceutiques." *Revue d'histoire de la pharmacie* 72, no. 262 (1984): 319–326.

Leveque, Cedric. "La gouvernementalité aux marges de l'Etat: La lutte contre le paludisme en Casamance (Sénégal)." PhD diss., Université de Bordeaux, 2015.

Lincoln, Sarah L. "Expensive Shit: Aesthetic Economies of Waste in Postcolonial Africa." PhD diss., Duke University, 2008.

Livingston, Julie. *Improvising Medicine: An African Oncology Ward in an Emerging Cancer Epidemic*. Durham, NC: Duke University Press, 2012.

Livsey, Tim. "'Suitable Lodgings for Students': Modern Space, Colonial Development and Decolonization in Nigeria." *Urban History* 41, no. 4 (2014): 664–685.

Lo, Yi-Chun, Carrie A. Dooyema, Antonio Neri, James Durant, Taran Jefferies, Andrew Medina-Marino, Lori de Ravello, Douglas Thoroughman, Lora Davis, Raymond S. Dankoll, Matthias Y. Samson, Ibrahima M. Luka, Ossai Okechukwu, Nasir T. Umar-Tsafe, Alhassan H. Dama, and Mary Jean Brown. "Childhood Lead Poisoning Associated with Gold Ore Processing: A Village-Level Investigation—Zamfara State, Nigeria, October–November 2010." *Environmental Health Perspectives* 120, no. 10 (2012): 1450–1455.

Loewenson, Rene. "Structural Adjustment and Health Policy in Africa." *International Journal of Health Services* 23, no. 4 (1993): 717–730.

Mamdani, Mahmood. *Scholars in the Marketplace: The Dilemmas of Neo-liberal Reform at Makerere University, 1989–2005*. Kampala: Fountain Publishers, 2007.

Marbury, Hugh J. "Hazardous Waste Exportation: The Global Manifestation of Environmental Racism." *Vanderbilt Journal of Transnational Law* 28 (1995): 251–296.

Markovitz, Irving Leonard. *Léopold Sédar Senghor and the Politics of Negritude*. New York: Atheneum, 1969.

Markowitz, Gerald, and David Rosner. *Deceit and Denial: The Deadly Politics of Industrial Pollution*. Berkeley: University of California Press, 2012.

Marks, Shula. *Divided Sisterhood: Race, Class and Gender in the South African Nursing Profession*. New York: St. Martin's Press, 1994.

Masquelier, Adeline Marie. "Behind the Dispensary's Prosperous Facade: Imagining the State in Rural Niger." *Public Culture* 13, no. 2 (2001): 267–291.

Masquelier, Adeline Marie. "Teatime: Boredom and the Temporalities of Young Men in Niger." *Africa* 83, no. 3 (2013): 470–491.

Mbembe, Achille. *On the Postcolony*. Berkeley: University of California Press, 2001.

Mbembe, Achille, and Janet Roitman. "Figures of the Subject in Times of Crisis." *Public Culture* 7, no. 2 (1995): 323–352.

Mbembe, Achille, and Libby Meintjes. "Necropolitics." *Public Culture* 15, no. 1 (2003): 11–40.

Means, Alex. "Toxic Sovereignty: Biopolitics and Côte d'Ivoire." *Politics and*

Culture, October 2, 2009. http://politicsandculture.org/2009/10/02/alex-means -toxic-sovereignty-biopolitics-and-cote-divoire/.

Meerman, I. Frans. *Mission au Projet FAO 'Effets sur l'environnement de la lutte antiacridienne (LOCUSTOX)': Rapport d'évaluation*. Wageningen, Netherlands: Wageningen Agricultural University, 1993.

Melly, Caroline. "Ethnography on the Road: Infrastructural Vision and the Unruly Present in Contemporary Dakar." *Africa* 83, no. 3 (2013): 385–402.

Melly, Caroline. "Inside-out Houses: Urban Belonging and Imagined Futures in Dakar, Senegal." *Comparative Studies in Society and History* 52, no. 1 (2010): 37–65.

Melrose, Dianna. *Bitter Pills: Medicines and the Third World Poor*. Oxford: Oxfam, 1982.

MGP-Afrique. "Étude de marché des prestations Projet 'Locustox.' Rapport Final." Dakar: MGP-Afrique, 1998.

Mika, Marissa. "Fifty Years of Creativity, Crisis, and Cancer in Uganda." *Canadian Journal of African Studies/Revue canadienne des études africaines* 50, no. 3 (2016): 395–413.

Mills, David. "Life on the Hill: Students and the Social History of Makerere." *Africa* 76, no. 2 (2006): 247–266.

Mitman, Gregg, Michelle Murphy, and Christopher Sellers, eds. *Landscapes of Exposure: Knowledge and Illness in Modern Environments*. Osiris, 2d ser., no. 19. Chicago: University of Chicago Press, 2004.

Mohamedbhai, Goolam. "The Effects of Massification on Higher Education in Africa." February 2008. http://ahero.uwc.ac.za/index.php?module=cshe&action =downloadfile&fileid=18409092513202791624126.

Monnais, Laurence. "Des poisons qui en disent long: Les fonctions de l'arsenal thérapeutique traditionnel du Viêtnam colonisé (1860–1954)." *Frontières* 16, no. 1 (2003): 12–19.

Monnais, Laurence, and Noémi Tousignant. "The Values of Versatility: Pharma- cists, Plants, and Place in the French (Post)Colonial World." *Comparative Studies in Society and History* 58, no. 2 (2016): 432–462.

Monosson, Emily. "Chemical Mixtures: Considering the Evolution of Toxicology and Chemical Assessment." *Environmental Health Perspectives* 113, no. 4 (2005): 383–390.

Monote, Serge Eloi. "Contribution à la détermination de la valeur marchande du lait en poudre commercialisé au Sénégal." Pharmacy thesis, Université Cheikh Anta Diop, Dakar, 1997.

Mourre, Martin. *Thiaroye 1944: Histoire et mémoire d'un massacre colonial*. Rennes, France: Presses Universitaires de Rennes, 2017.

Moyi Okwaro, Ferdinand, and P. Wenzel Geissler. "In/dependent Collaborations: Perceptions and Experiences of African Scientists in Transnational HIV Re- search." *Medical Anthropology Quarterly* 29, no. 4 (2015): 492–511.

Mulemi, Benson. "Technologies of Hope: Managing Cancer in a Kenyan Hospital." In *Making and Unmaking Public Health in Africa*, edited by Ruth Prince and Rebecca Marsland, 162–186. Athens: Ohio University Press, 2013.

Murphy, Michelle. *Sick Building Syndrome and the Problem of Uncertainty: En-*

vironmental Politics, Technoscience and Women Workers. Durham, NC: Duke University Press, 2006.

Nash, Linda Lorraine. *Inescapable Ecologies: A History of Environment, Disease, and Knowledge.* Berkeley: University of California Press, 2006.

Ndiaye, Ndeye Penda. "Contrôle de qualité des différentes marques de laits en poudre commercialisés au Sénégal." Pharmacy thesis, Université Cheikh Anta Diop, Dakar, 2002.

Ndour, Françoise. "Contrôle de certains métaux lourds dans les huitres de Joal-Fadiouth (Senegal) et incidence sur la santé publique." Pharmacy thesis, Université Cheikh Anta Diop, Dakar, 1989.

Ndour, Karim B. "Participation aux opérations menées par le projet LOCUSTOX: 'Effet sur l'environnement de la lutte anti-acridienne.'" Internship report, CILSS, Niamey, 1991.

Ndoye, Tidiane. *La société sénégalaise face au paludisme: Politiques, savoirs et acteurs.* Dakar: Éditions Karthala, 2008.

Nguyen, Vinh-Kim. "Antiretroviral Globalism, Biopolitics, and Therapeutic Citizenship." In *Global Assemblages: Technology, Politics, and Ethics as Anthropological Problems,* edited by Stephen Collier and Aiwa Ong, 124–144. Oxford: Blackwell, 2005.

Nguyen, Vinh-Kim. "Government-by-Exception: Enrolment and Experimentality in Mass HIV Treatment Programmes in Africa." *Social Theory and Health* 7, no. 3 (2009): 196–217.

Niane, Babacar. "Surveillance de la pollution marine au Sénégal: Analyse des métaux lourds dans les organismes marins d'importance commercial." Pharmacy thesis, Université Cheikh Anta Diop, Dakar, 1990.

Niane, Babacar, Amadou Diouf, B. Willer, Françoise Ndour, Mounirou Ciss, and Doudou Ba. "Surveillance de la pollution par les métaux lourds des huîtres de palétuviers du Sénégal." *Revue internationale d'océanographie médicale* 105–106 (1992): 71–78.

Niang, Souleymane. "De la politique scientifique senghorienne: Principes et stratégies." *Éthiopiques: Revue socialiste de culture négro-africaine.* Accessed January 23, 2014. http://ethiopiques.refer.sn/spip.php?article629.

Niassy, Abdoulaye, Alioune Beye, and Harold van Der Valk. "Impact of Fenitrothion Applications on Natural Mortality of Grasshopper Eggpods in Senegal (1991 Treatments)." In *Project Locustox: Environmental Side-Effects of Locust and Grasshopper Control.* Vol. 1, edited by James W. Everts, Djibril Mbaye, and Oumar Barry, 111–128. Dakar: FAO, LOCUSTOX Project, 1997.

Nixon, Rob. *Slow Violence and the Environmentalism of the Poor.* Cambridge, MA: Harvard University Press, 2011.

Nriagu, Jerome O. "Toxic Metal Pollution in Africa." *Science of the Total Environment* 121 (1992): 1–37.

Obi, Ejeatuluchukwu, Dora N. Akunyili, B. Ekpo, and Orish E. Orisakwe. "Heavy Metal Hazards of Nigerian Herbal Remedies." *Science of the Total Environment* 369, no. 1 (2006): 35–41.

O'Keefe, Phil. "Toxic Terrorism." *Review of African Political Economy* (1988): 84–90.

Okeke, Iruka N. "African Biomedical Scientists and the Promises of 'Big Science.'" *Canadian Journal of African Studies/Revue canadienne des études africaines* 50, no. 3 (2016): 455–478.

O'Neill, Bruce, and Dennis Rodgers, eds. "Special Issue on Infrastructural Violence." *Ethnography* 13, no. 4 (2012): 401–562.

Osseo-Asare, Abena Dove. "Bioprospecting and Resistance: Transforming Poisoned Arrows into Strophantin Pills in Colonial Gold Coast, 1885–1922." *Social History of Medicine* 21, no. 2 (2008): 269–290.

Osseo-Asare, Abena Dove. *Bitter Roots: The Search for Healing Plants in Africa.* Chicago: University of Chicago Press, 2014.

Osseo-Asare, Abena Dove. "Scientific Equity: Experiments in Laboratory Education in Ghana." *Isis* 104, no. 4 (2013): 713–741.

Patterson, Donna A. *Pharmacy in Senegal: Gender, Healing, and Entrepreneurship.* Bloomington: University of Indiana Press, 2015.

Patterson, Donna A. "Women Pharmacists in Twentieth-Century Senegal: Examining Access to Education and Property in West Africa." *Journal of Women's History* 24, no. 1 (2012): 111–137.

Patton, Adell. *Physicians, Colonial Racism and Diaspora in West Africa.* Gainesville: University Press of Florida, 1996.

Paye, Lucien. "Training Administrative Staff and Industrial and Commercial Workers in Africa." *Civilisations* 9, no. 3 (1959): 301–312.

Payet, Maurice, Gauthier Pille, Marc Sankale, P. Pene, and M. Trellu. "Le contexte humoral lipidique au cours de l'artherosclerose du Noir africain: Pathologie et biologie." *La semaine des hôpitaux* 9, nos. 9–10 (1961): 1093–1100.

Payet, Maurice, Gauthier Pille, P. Pene, and M. Moulanier. "A propos d'une intoxication volontaire par l'aspirine." *Bulletin de la Société médicale d'Afrique noire de langue française* 7, no. 3 (1962): 424–427.

Pazou, Elisabeth Yehouenou A., Michel Boko, Cornelis A. M. Van Gestel, Hyacinthe Ahissou, Philippe Lalèyè, Simon Akpona, Bert van Hattum, Kees Swart, and Nico M. van Straalen. "Organochlorine and Organophosphorous Pesticide Residues in the Ouémé River Catchment in the Republic of Bénin." *Environment International* 32, no. 5 (2006): 616–623.

Pellissier, Catherine. "Contribution à l'étude de la toxicologie analytique du chloralose Alpha." PhD diss., Université de Dakar, 1973.

Pellow, David Naguib. *Resisting Global Toxics: Transnational Movements for Environmental Justice.* Cambridge, MA: MIT Press, 2007.

Pesticides Referee Group. *Evaluation of Field Trial Data on the Efficacy and Selectivity of Insecticides on Locusts and Grasshoppers.* Report to FAO by the Pesticide Referee Group, sixth meeting, Rome, December 10–12, 1996. Accessed December 29, 2015. http://www.fao.org/ag/locusts/common/ecg/1155/en/PRG6e.pdf.

Peterson, Kristin. *Speculative Markets: Drug Circuits and Derivative Life in Nigeria.* Durham, NC: Duke University Press, 2014.

Petryna, Adriana. *Life Exposed: Biological Citizens after Chernobyl.* Princeton, NJ: Princeton University Press, 2013.

Pfeiffer, James. "International NGOs and Primary Health Care in Mozambique: The Need for a New Model of Collaboration." *Social Science and Medicine* 56, no. 4 (2003): 725–738.

Pfeiffer, James, and Rachel Chapman. "Anthropological Perspectives on Structural Adjustment and Public Health." *Annual Review of Anthropology* 39 (2010): 149–165.

Pille, Gauthier. "Conditions particulières à l'expertise toxicologique tropicale." *Bulletin de la Société médicale d'Afrique noire de langue française* 2 (1959): 158–164.

Pille, Gauthier. "Leçon inaugurale: [Faite le] 6 février 1963, Faculté mixte de médecine et de pharmacie de Dakar. Chaire de pharmacie chimique et de toxicologie." Excerpted from *Médecine d'Afrique noire*, 10, no. 4 (1963). Bibliothèque Interuniversitaire de Pharmacie, Faculty of Pharmacy of Paris, France.

Pille, Gauthier, M. Trellu, and Pauline Palancade. "Quelques aspects de la recherche chimique appliquée à la biologie et la pathologie dans le contexte climatique tropical." *Bulletins et mémoires de la Faculté nationale de médecine et de pharmacie de Dakar* 9 (1961): 326–336.

Pille, Gauthier, and Pauline Palancarde. "Quelques aspects actuels de l'expertise toxicologique à Dakar (importance des suicides par la chloroquine)." *Bulletins et mémoires de la Faculté mixte de médecine et de pharmacie de Dakar* 11 (1963): 264–275.

Piot, Charles. *Nostalgia for the Future: West Africa after the Cold War*. Chicago: University of Chicago Press, 2010.

Pollock, Anne. "Places of Pharmaceutical Knowledge-Making: Global Health, Postcolonial Science, and Hope in South African Drug Discovery." *Social Studies of Science* 44, no. 6 (2014): 848–873.

Portmann, J. E., C. Biney, A. C. Ibe, and S. Zabi. *State of the Marine Environment: West and Central African Region*. UNEP Regional Seas Reports and Studies no. 108, UNEP, 1989. Accessed April 3, 2017. http://wedocs.unep.org/bitstream /handle/20.500.11822/11729/rsrs108.pdf?sequence=1&isAllowed=y.

Povinelli, Elizabeth A. "The Woman on the Other Side of the Wall: Archiving the Otherwise in Postcolonial Digital Archives." *Differences* 22, no. 1 (2011): 146–171.

Prakash, Gyan. *Another Reason: Science and the Imagination of Modern India*. Princeton, NJ: Princeton University Press, 1999.

Prince, Ruth. "Situating Health and the Public in Africa: Historical and Anthropological Perspectives." In *Making and Unmaking Public Health in Africa*, edited by Ruth Prince and Rebecca Marsland, 1–54. Athens: Ohio University Press, 2013.

Quet, Mathieu. "Sécurisation pharmaceutique et économies du medicament: Controverses globales autour des politiques anti-contrefaçon." *Sciences sociales et santé* 33, no. 1 (2015): 91–116.

Rambourg-Schepens, Marie-Odile, Antoine Pitti-Ferrandi, Martine Ledrans, and Michel Jouan. "Rapport de la mission InVS/Afsse sur les Centres antipoison et les Centres de toxicovigilance." September 2003. http://opac.invs.sante.fr/doc _num.php?explnum_id=5514.

Rattner, Barnett A. "History of Wildlife Toxicology." *Ecotoxicology* 18, no. 7 (2009): 773–783.

Redfield, Peter. "Bioexpectations: Life Technologies as Humanitarian Goods." *Public Culture* 24 (2012): 157–184.

Redfield, Peter. "The Half-Life of Empire in Outer Space." *Social Studies of Science* 32, no. 5–6 (2002): 791–825.

Redfield, Peter. *Life in Crisis: The Ethical Journey of Doctors without Borders.* Berkeley: University of California Press, 2013.

Rees, Tobias. "Humanity/Plan; or, on the 'Stateless' Today (Also Being an Anthropology of Global Health)." *Cultural Anthropology* 29, no. 3 (2014): 457–478.

Reid-Henry, Simon M. *The Cuban Cure: Reason and Resistance in Global Science.* Chicago: University of Chicago Press, 2010.

République du Sénégal. "Quatre années de fonctionnement de la Délégation Generale à la Recherche Scientifique et Technique: Le point sur la politique scientifique et technique du Sénégal et les perspectives de son evolution." Dakar: Janvier 1978. National Library of Senegal, Dakar.

République du Sénégal, Ministère de la santé et de la prevention. "Plan national de développement sanitaire: PNDS 2009–2018." January 2009. https://www.internationalhealthpartnership.net/fileadmin/uploads/ihp/Documents/Country_Pages/Senegal/PNDS2009_2018.pdf.

Ridde, Valéry. "Per Diems Undermine Health Interventions, Systems and Research in Africa: Burying our Heads in the Sand." *Tropical Medicine and International Health* (2010): doi: 10.1111/j.1365–3156.2010.02607.x.

Riddell, J. Barry. "Things Fall Apart Again: Structural Adjustment Programmes in Sub-Saharan Africa." *Journal of Modern African Studies* 30, no. 1 (1992): 53–68.

Robbins, Bruce. "The Smell of Infrastructure: Notes toward an Archive." *boundary* 2 34, no. 1 (2007): 25–33.

Roberts, Jody A., and Nancy Langston, eds. "Toxic Bodies/Toxic Environments: An Interdisciplinary Forum." *Environmental History* 13, no. 4 (2008): 629–703.

Rocheteau, Guy, Jean Roch, and Philippe Hugon. *Pouvoir financier et indépendance économique en Afrique: Le cas du Sénégal.* Paris: Karthala-ORSTOM, 1982.

Roy, Jean. *Histoire d'un siècle de lutte anti-acridienne en Afrique: Contributions de la France.* Paris: Editions l'Harmattan, 2001.

Sabatier, Peggy R. "'Elite' Education in French West Africa: The Era of Limits, 1903–1945." *International Journal of African Historical Studies* 11, no. 2 (1978): 247–266.

Saine, Joseph Emmanuel. "Aflatoxines dans l'huile de pression artisanale d'arachide: Essais de décontamination par filtration sur du charbon de bois." Pharmacy thesis, Université Cheikh Anta Diop, Dakar, 2001.

Sall, Amadou. "Recherche et dosage des aflatoxines dans les pâtes d'arachide alimentaires." Pharmacy thesis, Université Cheikh Anta Diop, Dakar, 1998.

Samou, Jackson Edouard. "Les aliments enrichis de vitamine D: Contrôle analytique d'échantillons de lait en poudre au Sénégal." Pharmacy thesis, Université Cheikh Anta Diop, Dakar, 1995.

Sankalé, Marc, Louis-Vincent Thomas, and Pierre Fougeyrollas, eds. *Dakar en devenir.* Dakar: Présence africaine, 1968.

Sarr, Ibrahima. "Détermination de l'impact potentiel des pesticides sur Heliochei-lus albipunctella (mineuse de l'épi du mil), à partir d'une méthode indirecte, l'étude de la table de survie." Agronomic engineering diploma thesis, École Nationale Supérieure d'Agriculture de Thiès, 1996.

Sarr, Mouhamadou Moctar. "Contribution à l'étude de la pollution atmosphérique due au plomb émis par les véhicules dans la ville de Dakar." Pharmacy thesis, Université Cheikh Anta Diop, Dakar, 1992.

Schaffer, Simon. "Easily Cracked: Scientific Instruments in States of Disrepair." *Isis* 102, no. 4 (2011): 706–717.

Schumacher, Edward J. *Politics, Bureaucracy, and Rural Development in Senegal.* Berkeley: University of California Press, 1975.

Sellers, Christopher C. "Cross-Nationalizing the History of Industrial Hazard." *Medical History* 54, no. 3 (2010): 315–340.

Sellers, Christopher C. *Hazards of the Job: From Industrial Disease to Environmental Health Science.* Chapel Hill: University of North Carolina Press, 1997.

Sellers, Christopher C., and Joseph Melling, eds. *Dangerous Trade: Histories of Industrial Hazard across a Globalizing World.* Philadelphia: Temple University Press, 2011.

Shapin, Steven. "The Invisible Technician." *American Scientist* 77, no. 6 (1989): 554–563.

Shapin, Steven. *The Scientific Life: A Moral History of a Late Modern Vocation.* Chicago: University of Chicago Press, 2009.

Sidia, Abdallahi Ould M., R. Skaf, J. M. Castel, A. Ndiaye, and J. A. Whellan. "OCLALAV and Its Environment: A Regional International Organization for the Control of Migrant Pests [and Discussion]." *Philosophical Transactions of the Royal Society of London. B, Biological Sciences* 287, no. 1022 (1979): 269–276.

Simone, AbdouMaliq. "People as Infrastructure: Intersecting Fragments in Johannesburg." *Public Culture* 16, no. 3 (2004): 407–429.

Sobukola, Olajide Philip, O. M. Adeniran, A. A. Odedairo, and O. E. Kajihausa. "Heavy Metal Levels of Some Fruits and Leafy Vegetables from Selected Markets in Lagos, Nigeria." *African Journal of Food Science* 4, no. 2 (2010): 389–393.

Soske, Jon. "'The Dissimulation of Race': Afro-Pessimism and the Problem of Development." *Qui Parle* 14, no. 2 (2004): 15–56.

Stoler, Ann Laura. *Along the Archival Grain: Epistemic Anxieties and Colonial Common Sense.* Princeton, NJ: Princeton University Press, 2010.

Stoler, Ann Laura. "Imperial Debris: Reflections on Ruins and Ruination." *Cultural Anthropology* 23, no. 2 (2008): 191–219.

Sullivan, Noelle. "Mediating Abundance and Scarcity: Implementing an HIV/AIDS-Targeted Project within a Government Hospital in Tanzania." *Medical Anthropology* 30, no. 2 (2011): 202–221.

Sy, Fatimata Oumar. "Contribution à l'étude du thé vert de chine utilisé en Afrique de l'Ouest: Contrôle de quelques éléments toxicologiques et incidences sur la santé publique." Pharmacy thesis, Université Cheikh Anta Diop, Dakar, 1991.

Tadjo, Véronique. "Dessine-moi (écris-moi) une independence. . . ." In *African Re-*

naissances Africaines: Writing 50 Years of Independence/Ecrire 50 ans d'indépendaence, edited by Gustave Akapko, 66–67. Milan: Silvana Editoriale, 2010.

Tagwireyi, Dexter, Douglas E. Ball, and Charles F. B. Nhachi. "Poisoning in Zimbabwe: A Survey of Eight Major Referral Hospitals." *Journal of Applied Toxicology* 22, no. 2 (2002): 99–105.

TAMS Consultants and Consortium for International Crop Protection (CICP). *Locust and Grasshopper Control in Africa/Asia: A Programmatic Environmental Assessment*. Washington, DC: USAID, 1989.

Thiam, Abderrahmane. "Elaboration de la table de survie partielle de Heliocheilus albipunctella (De Joannis)." Agricultural engineering diploma, Ecole Nationale des Cadres Ruraux, Bambey, 1995.

Third World Network. *Toxic Terror*. Penang: Third World Network, 1989.

Thompson, Edward P. "Time, Work-Discipline, and Industrial Capitalism." *Past and Present*, no. 38 (December 1967): 56–97.

Thorpe, Charles. "Against Time: Scheduling, Momentum, and Moral Order at Wartime Los Alamos." *Journal of Historical Sociology* 17, no. 1 (2004): 31–55.

Tichenor, Marlee. "The Power of Data: Global Health Citizenship and the Senegalese Data Retention Strike." In *Metrics: What Counts in Global Health*, edited by Vincanne Adams, 105–124. Durham, NC: Duke University Press, 2016.

Timmermans, Stefan. "A Black Technician and Blue Babies." *Social Studies of Science* 33, no. 2 (2003): 197–229.

Timmermans, Stefan, and Steven Epstein. "A World of Standards but Not a Standard World: Toward a Sociology of Standards and Standardization." *Annual Review of Sociology* 36 (2010): 69–89.

Touré, Kamadore, Malang Coly, Dieynaba Toure, Maouly Fall, Moussa Deng Sarr, Amadou Diouf, Mame Demba Sy, Mathias Camara, Fatoumata Diène Sarr, Joseph Faye, Gaoussou Diakhaby, Abdoulaye Badiane, Njido Ardo Bar, Amadou Gallo Diop, Mouhamadou Mansour Ndiaye, the Investigation Team, and Adama Tall. "Investigation of Death Cases by Pesticides Poisoning in a Rural Community, Bignona, Senegal." *Epidemiology: Open Access* 1, no. 2 (2011). http://dx.doi.org/10.4172/2161-1165.1000105.

Tourte, René, and M. Le Moigne. "L'équipement rural au Sénégal: Rôle de la recherche agronomique et de sa division du machinisme agricole et génie rural." *Machinisme agricole tropical* 31 (1970): 3–17.

Tousignant, Noémi. "Broken Tempos: Of Means and Memory in a Senegalese University Laboratory." *Social Studies of Science* 43, no. 5 (2013): 729–753.

Tousignant, Noémi. "Half-built Ruins." In *Traces of the Future: An Archaeology of Medical Science in Twenty-First Century Africa*, edited by P. Wenzel Geissler, Guillaume Lachenal, John Manton, and Noémi Tousignant, 35–38. Bristol, UK: Intellect Press, 2016.

Tousignant, Noémi. "Insects-as-Infrastructure: Indicating, Project Locustox and the Sahelization of Ecotoxicology." *Science as Culture* 22, no. 1 (2013): 108–131.

Tousignant, Noémi. "Pharmacy, Money and Public Health in Dakar." *Africa* 83, no. 4 (2013): 561–581.

Tousignant, Noémi. "The Qualities of Citizenship: Private Pharmacists and the State in Senegal after Independence and Alternance." In *Making and Unmaking Public Health in Africa*, edited by Ruth Prince and Rebecca Marsland, 96–115. Athens: Ohio University Press, 2013.

Tuakuila, Joel, Martin Kabamba, Honoré Mata, and Gerard Mata. "Blood Lead Levels in Children after Phase-out of Leaded Gasoline in Kinshasa, the Capital of Democratic Republic of Congo (DRC)." *Archives of Public Health* 71, no. 1 (2013). doi: 10.1186/0778-7367-71-5.

Turshen, Meredith. *Privatizing Health Services in Africa*. New Brunswick, NJ: Rutgers University Press, 1999.

Uduku, Ola. "Modernist Architecture and 'the Tropical' in West Africa: The Tropical Architecture Movement in West Africa, 1948–1970." *Habitat International* 30, no. 3 (2006): 396–411.

United Nations Environmental Program (UNEP). "First Workshop of Participants in the Joint FAO-IOC-WHO-IAEA-UNEP Project on Monitoring of Pollution in the Marine Environment of the West and Central African Region (WACAF/2—Pilot Phase)." Intergovernmental Oceanographic Commission, Workshop report no. 41. Dakar, Senegal, October 28–November 1, 1985. Accessed April 5, 2017. http://unesdoc.unesco.org/images/0007/000721/072120e0.pdf.

United Nations Environmental Program (UNEP). "The Hazardous Chemicals and Wastes Conventions." October 2004. http://archive.basel.int/pub/three Conventions.pdf.

United Nations Environmental Program (UNEP). "The West and Central African Action Plan: Evaluation of Its Developments and Achievements." UNEP Regional Seas Reports and Studies no. 101, UNEP, 1989.

United Nations Environment Program (UNEP) and Njoroge G. Kimani. "Environmental Pollution and Impacts on Public Health: Implications of the Dandora Municipal Dumping Site in Nairobi, Kenya, Report Summary." Accessed April 4, 2017. http://staging.unep.org/urban_environment/PDFs/Dandora WasteDump-ReportSummary.pdf.

United Nations Institute for Training and Research (UNITAR). "UNITAR/IOMC Programme to Assist Countries in Developing and Sustaining an Integrated National Programme for the Sound Management of Chemicals: 1999–2002." Programme Document, January 2000. http://cwm.unitar.org/publications /publications/cw/inp/inp99-02_(09_jan_02).pdf. http://cwm.unitar.org /national-profiles/publications/cw/inp/inp99-02_(09_jan_02).pdf.

United States Agency for International Development (USAID). *Review of Environmental Concerns in A.I.D. Programs for Locust and Grasshopper Control*. Washington, DC: USAID, 1991.

United States Pharmacopeial Convention (USP). "DQI Proposed Work Plan, Senegal, October 1, 2008–September 30, 2009." Accessed August 13, 2015. http://www .pmi.gov/docs/default-source/default-document-library/implementing-partner -reports/dqi-proposed-work-plan-senegal-fy2009.pdf?sfvrsn=4.

United States Pharmacopeial Convention (USP). "Promoting the Quality of Medicines in Developing Countries." Accessed August 13, 2015. http://www.usp.org /global-health-programs/promoting-quality-medicines-pqmusaid.

Vaillant, Janet G. "Perspectives on Leopold Senghor and the Changing Face of Negritude. Review of *The Concept of Negritude in the Poetry of Leopold Sedar Senghor*, by Sylvia Washington Bâ, Leopold Sedar Senghor; *Leopold Sedar Senghor: An Intellectual Biography*, by Jacques-Louis Hymans; *Leopold Sedar Senghor and the Politics of Negritude*, by Irving Markovitz; *Leopold Sedar Senghor et la naissance de l'Afrique moderne*, by Ernest Milcent, Monique Sordet." *ASA Review of Books* 2 (1976): 154–162.

van der Stoep, Jan. "The Development of Laboratory Toxicity Tests with Bracon Hebetor (Say) for the Study of the Environmental Effects of Locust Control: An Ecotoxicological Study of the Effects of Fenitrothion and Diflubenzuron on Population Dynamics Parameters of Bracon Hebetor." Doctoral research report, Wageningen Agricultural University, n.d.

van der Valk, Harold, Jan van der Stoep, Baba Fall, and Eloi Dieme. "A Laboratory Toxicity Test with Bracon Hebetor (Say) (Hymenoptera, Braconidae): First Evaluation of Rearing and Testing Methods." In *Locustox Project: Environmental Side-Effects of Locust and Grasshopper Control, Vol. I*, edited by James W. Everts, Djibril Mbaye, and Wim C. Mullié, 123–154. Rome: FAO, 1997.

van der Valk, Harold, and Ousmane Kamara. "The Effect of Fenitrothion and Diflubenzuron on Natural Enemies of Millet Pests in Senegal (the 1991 Study)." In *Locustox Project: Environmental Side-Effects of Locust and Grasshopper Control, Vol. I*, edited by James W. Everts, Djibril Mbaye, and Wim C. Mullié, 61–106. Rome: FAO, 1997.

Van Dyk, J. Susan, and Brett Pletschke. "Review on the Use of Enzymes for the Detection of Organochlorine, Organophosphate and Carbamate Pesticides in the Environment." *Chemosphere* 82, no. 3 (2011): 291–307.

Van Straaten, Peter. "Mercury Contamination Associated with Small-scale Gold Mining in Tanzania and Zimbabwe." *Science of the Total Environment* 259, no. 1 (2000): 105–113.

Vogel, Sarah A. "Forum: From 'The Dose Makes the Poison' to 'The Timing Makes the Poison': Conceptualizing Risk in the Synthetic Age." *Environmental History* 13, no. 4 (2008): 667–673.

Vogel, Sarah A. *Is it Safe? BPA and the Struggle to Define the Safety of Chemicals*. Oakland, CA: University of California Press, 2012.

Von Schnitzler, Antina. "Citizenship Prepaid: Water, Calculability, and Techno-Politics in South Africa." *Journal of Southern African Studies* 34, no. 4 (2008): 899–917.

Waast, Roland. "L'état des sciences en Afrique: Vue d'ensemble." Direction Générale de la Coopération Internationale et du Développement, Ministère des Affaires Étrangères, April 2002.

Weir, David. *The Bhopal Syndrome: Pesticides, Environment, and Health*. San Francisco: Sierra Club Books, 1987.

Weir, David, and Mark Schapiro. *Circle of Poison: Pesticides and People in a Hungry World*. Oakland, CA: Institute for Food and Development Policy, 1981.

Wendland, Claire L. *A Heart for the Work: Journeys through an African Medical School*. Chicago: University of Chicago Press, 2010.

Wendland, Claire L. "Moral Maps and Medical Imaginaries: Clinical Tourism at Malawi's College of Medicine." *American Anthropologist* 114, no. 1 (2012): 108–122.

Werbner, Richard, ed. *Memory and the Postcolony: African Anthropology and the Critique of Power*. London: Zed Books, 1998.

Whyte, Susan Reynolds. "Knowing Hypertension and Diabetes: Conditions of Treatability in Uganda." *Health and Place* 39 (2016): 219–225.

Whyte, Susan Reynolds. "Pharmaceuticals as Folk Medicine: Transformations in the Social Relations of Health Care in Uganda." *Culture, Medicine and Psychiatry* 16, no. 2 (1992): 163–186.

Whyte, Susan Reynolds. "The Publics of the New Public Health: Life Conditions and Lifestyle Diseases in Uganda." In *Making and Unmaking Public Health in Africa,* edited by Ruth Prince and Rebecca Marsland, 187–207. Athens: Ohio University Press, 2013.

Whyte, Susan Reynolds, Michael Whyte, Lotte Meinert, and Jenipher Twebaze. "Therapeutic Clientship: Belonging in Uganda's Projectified Landscape of AIDS Care." In *When People Come First: Critical Studies in Global Health*, edited by Joao Biehl and Adriana Petryna, 140–165. Princeton, NJ: Princeton University Press, 2013.

Wilder, Gary. *Freedom Time: Negritude, Decolonization, and the Future of the World*. Durham, NC: Duke University Press, 2014.

Williams, Jonathan H., Timothy D. Phillips, Pauline E. Jolly, Jonathan K. Stiles, Curtis M. Jolly, and Deepak Aggarwal. "Human Aflatoxicosis in Developing Countries: A Review of Toxicology, Exposure, Potential Health Consequences, and Interventions." *American Journal of Clinical Nutrition* 80, no. 5 (2004): 1106–1122.

Wilson, Brian. "Technical Expert Mission in Senegal to Determine the Best Approach for the Introduction of Environmentally Sound Procedures for the Recovery of Used Lead Acid Batteries in a Manner Consistent with the Basel Convention Technical Guidelines. Dakar, Senegal, 21–27 April 2008." Report prepared for the Secretariat of the Basel Convention by the ILMC Expert, May 29, 2008. http://www.ilmc.org/Basel%20Project/Senegal/Reports/Technical%20Mission%20Report%20for%20Senegal%2029%20May%202008.pdf.

World Health Organization. *Guidelines for Poison Control*. Geneva: WHO, 1997.

World Health Organization. "Intoxication au plomb à Thiaroye sur mer, Sénégal. Mission d'appui de l'OMS, 7–21 juin 2008." Accessed October 20, 2017. http://www.who.int/environmental_health_emergencies/events/french%20page%202.pdf?ua=1.

Yabe, John, Mayuni Ishizuka, and Takashi Umemura. "Current Levels of Heavy Metal Pollution in Africa." *Journal of Veterinary Medical Science* 72, no. 10 (2010): 1257–1263.